# THEORETICAL
# NUMERICAL ANALYSIS

# THEORETICAL NUMERICAL ANALYSIS

## An Introduction to Advanced Techniques

**PETER LINZ**

*University of California*
*Davis, California*

A WILEY-INTERSCIENCE PUBLICATION

**JOHN WILEY & SONS**

New York • Chichester • Brisbane • Toronto • Singapore

**Library of Congress Cataloging in Publication Data**

Linz, Peter.
  Theoretical numerical analysis.

  (Pure and applied mathematics)
  "A Wiley-Interscience publication."
  Includes index.
  1. Numerical analysis. I. Title.
QA297.L55   1978          519.4                78-15178
ISBN 0-471-04561-6

Printed in the United States of America

10 9 8 7 6 5 4 3 2

# PREFACE

While the primary intent of this book is to serve as a text for a one-year graduate course in theoretical numerical analysis, it was also written to provide engineers and scientists experienced in numerical computing with a simple introduction to the major ideas of modern numerical analysis. It assumes that the reader has some practical experience with computational mathematics and the ability to relate this experience to new concepts. Otherwise, no mathematical background beyond advanced calculus is assumed. The ideas of functional analysis that are used throughout are introduced and developed to the extent to which they are needed.

The impetus for writing the book came from a feeling of dissatisfaction I have about the way in which numerical analysis is usually taught. Certainly in undergraduate courses and frequently even at the graduate level there is a strong emphasis on special techniques, clever tricks, and ad hoc procedures which often obscure the essential nature of the subject and make it appear as separate from traditional mathematics. The student may be left with the idea that numerical analysis consists of a bag of tricks which, when used with enough ingenuity, will yield the solution of almost any problem. This attitude contrasts sharply with that of the theoretical numerical analyst who has a tendency to produce highly abstract theories that have little apparent connection with actual problems and are written in a language unfamiliar to most nonmathematicians. As a result much of the useful material produced by researchers in numerical analysis is not easily accessible to engineers and others needing to solve real problems. This situation is unfortunate and should be remedied. I hope that this book will serve to close the communications gap at least somewhat.

Since the advent of the digital computer, numerical analysis has become one of the most important and frequently used tools of scientists and engineers faced with the necessity of having to solve complex mathematical problems. Some employ numerical techniques routinely, while others may consider them "methods of last resort" to be used only when everything else fails. Be that as it may, the success of numerical analysis in solving otherwise intractable problems is beyond dispute. This widespread use of numerical methods has come about not only because of the failure of classical mathematical analysis but also because, in some sense, numerical analysis is easy. By this I mean that it is often possible for someone with little more than a knowledge of undergraduate mathematics to design algorithms for the approximate solution of rather difficult problems and so write computer programs that, on the surface at least, perform quite adequately. This is not to say that producing such a program is always a trivial process; most of the time one has to use a good deal of physical intuition and ingenuity to make it work. Nevertheless, a deep knowledge of mathematics is often not required. As a result there exists among practically oriented scientists a certain disdain for the theoretical aspects of numerical analysis. The feeling that what is intuitively reasonable will work and that theorems contribute little seems to be widespread. Mathematicians have reacted to this by claiming that such an intuitive approach is fraught with danger and should not be dignified with the name analysis or mathematics, but should rather be called "numerical experimentation." Some numerical analysts seem to have overreacted, making their work more and more abstract and sacrificing intelligibility for the sake of generality, rigor, and mathematical elegance. Even a cursory glance at what is currently being published will bear out the validity of this observation.

This dispute has gone on for some time and undoubtedly will continue to do so. If I were forced to choose I would very likely side with the engineer; numerical analysis is after all a practical discipline in which problem-solving is more important than theorem-proving. However, having started with a bias toward the practical side, I have gradually come to appreciate the power of the theoretical approach and have become convinced that it is useful for anyone doing numerical computations to have some understanding of the mathematical foundations of numerical analysis. Remembering Richard Hamming's classic dictum that the purpose of computing is insight, not numbers, we ought to consider a theorem not as an end in itself, but as something which clarifies the underlying ideas on which a specific algorithm is based. Good problem-solving requires above all insight and understanding; such understanding comes not only from experience but also from an ability to see a problem in a wider context. Finding a method that will solve a given problem is often not hard, but producing an efficient, well-constructed algorithm whose behavior is predictable and whose results can be judged by more stringent rules than mere plausibility is much more challenging. To those who are not satisfied with just computing but want to compute well, theoretical numerical analysis can be of great help.

The notion that understanding and insight are the most important purposes of a theoretical treatment of numerical analysis has been the major factor which determined my choice of content and style of presentation. The material has been arranged to be read in the sequence in which it is presented and each concept should be well understood before proceeding. The book is not to be considered as a reference text for theorems in numerical analysis; many of the results are valid only in the context of a particular section and the reader should be careful not to use them out of this context. I have tried to state results as simply as possible by limiting the generality and avoiding excessive mathematical shorthand notation. Even a moderately sophisticated reader may discover obvious extensions; the fact that they are not explicitly presented should not be taken as an indication that such generalizations are unknown. I have given proofs that I felt were not too technical and that contributed something to the understanding. Proofs that are not of this nature and that can easily be found elsewhere have been omitted. What I consider important is the content of a theorem, not how it is proved. For example, one needs to know what the Weierstrass approximation theorem says, but the proof is probably of little interest to anyone except a specialist in approximation theory. Some proofs, which are easy extensions of what has preceded, have been relegated to the exercises and can be used by the reader as a test of his understanding. In general, the exercises must be considered as an integral part of the presentation and should all be done carefully. Overall, my aim was to be as concise as possible; consequently the reader is frequently referred to the literature for details.

The material has been divided into three main parts. Part I presents some of the fundamental notions of functional analysis and approximation theory. To the mathematically trained reader much of this will undoubtedly be familiar; nevertheless it should be read carefully. In the least it will establish the notation and single out the concepts which are important in the rest of the book. Part II outlines the major results of theoretical numerical analysis. Here are to be found the central ideas which every numerical analyst should know. Finally, for Part III I have selected some special topics to show the power and usefulness of the theoretical analysis. Much of this is of fairly recent origin and contains a number of open problems for those interested in doing research in this area.

The book is the outgrowth of a one-year graduate course in numerical analysis which I have taught at the University of California at Davis for a number of years. It has been my experience that all of the material can be covered in somewhat less than a year, leaving room for some supplementation. How this is to be done must of course be left to the taste of the instructor, but some particular suggestions may be helpful. If one were to present this material to mathematics students with little practical experience one might assign a number of computer programming exercises (perhaps centered around the ideas in Part II) and ask that the student

viii PREFACE

present their algorithms and discuss the results in class. In other cases, the instructor may have some special area of interest which he wishes to explore in more detail. My own approach has been to cover the whole book (which I normally do in 60 to 65 lecture hours), then ask students to pick a specific topic, read some research papers on it, and present their findings to the rest of the class. This works well in a situation where most of the students are prospective Ph.D. candidates.

Over the years this work has gradually evolved, often through trial-and-error, to its present form. This development has greatly benefitted from the questions, criticisms, independent insights, and occasional difficulties of a number of my students. To all of those who participated in this process I wish to express my gratitude.

PETER LINZ

*Davis, California*
*August 1978*

# CONTENTS

# PART I

# BASIC CONCEPTS

Numerical analysis or, as it is perhaps more appropriately called, computational mathematics is the study of methods for the numerical (and, in general, approximate) solution of mathematically posed problems. Loosely speaking one can distinguish between two different aspects of computational mathematics: *methodology*, which deals with the construction of specific algorithms, their efficiency, implementation for computers, and various other practical questions, and *analysis*, in which one studies the underlying principles, error bounds, convergence theorems, and so on. It is this second aspect that is our main concern.

Most introductory treatments of numerical analysis stress methodology, and the student is exposed to a number of different algorithms which, on the surface, seem to have little in common. A deeper study, however, reveals a different picture. There are a few basic principles on which most of computational mathematics is based; often seemingly different methods are based on the same general idea. It is our aim to study these basic principles and to single out the fundamental notions which are the heart of modern numerical analysis. No attempt will be made to discuss particular problems or algorithms, except as examples for a general theory. To understand the principles on which numerical analysis is based is not only of interest to the mathematician, but can also be of considerable help in practical computation since a clear idea of the fundamentals is valuable in the construction of solution methods for the difficult problems encountered in practice.

We will begin by reviewing some simple results from functional analysis and approximation theory. A knowledge of these concepts is indispensable for further work; functional analysis provides us with a language for the development of a general theory, while approximation theory makes it possible to relate this theory to specific cases in a meaningful way. Both subjects are of course very extensive and contain topics of little use to the numerical analyst. Fortunately, the ideas needed here are the simpler ones, so that we need not delve into these subjects very deeply.

# 1
# REVIEW OF
# FUNCTIONAL ANALYSIS

Modern numerical analysis has benefited considerably from its close association with functional analysis, which provides it not only with an elegant and concise notation, but also with an efficient tool for the development of new theories. Because of this, it is becoming increasingly more difficult for anyone unfamiliar with functional analysis to appreciate the current developments in numerical analysis. Fortunately, matters are not as difficult as they sometimes seem and it is not necessary to become an expert in functional analysis; what we mainly need here is a working knowledge of some quite simple concepts. This chapter is devoted to a review of the most basic ideas of functional analysis which, for the most part, are sufficient for further discussions. Occasionally, particularly in later chapters, a few more sophisticated concepts will be helpful, but these will be introduced when needed.

## 1.1 LINEAR SPACES

**DEFINITION 1.1.** Let $X = \{x, y, z, \ldots\}$ be a set and $F = \{\alpha, \beta, \gamma, \ldots\}$ a scalar field. Let there be defined an operation of *addition* between every two elements of $X$, and an operation of *scalar multiplication* between every element of $F$ and every element of $X$ such that

(a) If $x$ is in $X$ and $y$ is in $X$ then $x + y$ is in $X$. In mathematical shorthand notation we will write this as $x \in X, y \in X \Rightarrow x + y \in X$.

(b) $x \in X, \alpha \in F \Rightarrow \alpha x \in X$.

(c) $x + y = y + x$.

(d) $(x + y) + z = x + (y + z)$.

(e)  There exists an element $0 \in X$ such that $x + 0 = x$ for all $x \in X$.*
(f)  For each $x \in X$ there exists a unique element, called the negative of $x$ and denoted by $-x$, such that $x + (-x) = 0$.
(g)  $\alpha(\beta x) = (\alpha\beta)x$.
(h)  $\alpha(x + y) = \alpha x + \alpha y$.
(i)  $(\alpha + \beta)x = \alpha x + \beta x$.
(j)  $1x = x$.

Then $X$ is called a linear space over the field $F$. If $F$ is the field of real numbers, $X$ is a *real linear space*; if $F$ is the field of complex numbers, $X$ is a *complex linear space*. We normally deal only with real spaces and, unless otherwise stated, we use the term linear space to denote a real linear space.

**Example 1.1.**  There are several familiar examples of linear spaces which play an important role in applied mathematics and numerical analysis:

(a)  The set of all vectors with $n$ components, with the usual definition of addition and scalar multiplication, is a linear space. We denote this space by $R^n$. Much of the terminology for the general case has been adopted from this example. Thus a linear space is often called a vector space and its elements are referred to as vectors or points.
(b)  The set of all continuous functions on some interval $[a,b]$ is a linear space, denoted by $C[a,b]$. Again, addition and scalar multiplication are defined in the usual way, that is, pointwise.
(c)  The set of all $n$-times continuously differentiable functions on $[a,b]$ is a linear space, denoted by $C^{(n)}[a,b]$.
(d)  The set of all functions $x(t)$ for which

$$\int_R |x(t)|^p \, dt$$

exists is the linear space $L_p$ or $L_p[R]$. The integral here is to be interpreted in the Lebesgue sense, but the reader who is unfamiliar with Lebesgue integration may, without much loss, think of it as a Riemann integral. Of particular interest in mathematics is $L_2$, the space of all square-integrable functions.

---

*The 0 here denotes the zero element in $X$ and should, strictly speaking, be distinguished from the zero element of $F$. But since the context always removes any possible ambiguity we use the same symbol for both zeros.

(e)   The set of all infinite sequences $x_1, x_2, \ldots$ of real numbers such that

$$\sum_{i=1}^{\infty} |x_i|^p < \infty$$

is a linear space, denoted by $l_p$.

**DEFINITION 1.2.**   Let $x_1, x_2, \ldots$ be elements of a linear space, and $\alpha_1, \alpha_2, \ldots$ be elements of the underlying field. The sum

$$\alpha_1 x_1 + \alpha_2 x_2 + \cdots + \alpha_n x_n$$

is called a *linear combination* of the $x_i$. The elements $x_i$ are said to be *linearly independent* if and only if

$$\alpha_1 x_1 + \alpha_2 x_2 + \cdots + \alpha_n x_n = 0 \Rightarrow \alpha_1 = \alpha_2 = \cdots = \alpha_n = 0.$$

Otherwise, if there exists a set $\alpha_1, \alpha_2, \ldots, \alpha_n$, not all zero, such that the linear combination is zero, the set is said to be *linearly dependent*.

**DEFINITION 1.3.**   If there exist elements $x_1, x_2, \ldots, x_n \in X$ which are linearly independent, but every set of $n + 1$ elements is linearly dependent, then we call $n$ the *dimension* of $X$, denoted by $\dim(X)$. If for every $n > 0$ there exist $n$ linearly independent elements in $X$, then $X$ is said to be of *infinite* dimension.

**DEFINITION 1.4.**   A linearly independent set $x_1, x_2, \ldots, x_n \in X$ is said to be a *basis* for $X$ if every $x \in X$ can be expressed as a linear combination of the $x_i$.

This definition of a basis is appropriate when $X$ is of finite dimension; for infinite-dimensional spaces the analogous definition requires the notion of a convergent series and will have to be deferred until the next section.

**THEOREM 1.1.**   A linear space $X$ is of finite dimension $n$ if and only if it has a basis of $n$ elements. Also, any set of $n$ linearly independent elements $x_1, x_2, \ldots, x_n$ constitutes a basis for an $n$-dimensional space, called the *span* of $x_1, \ldots, x_n$.

The proof is left as an exercise.

**Example 1.2**

(a)   The set of all vectors with $n$ components forms an $n$-dimensional space.

(b) The set of all polynomials of degree not greater than $n$ is a space $\mathscr{P}_n$ of dimension $n+1$. The powers $1, t, t^2, \ldots, t^n$ are a basis for $\mathscr{P}_n$.

(c) The space of continuous functions $C[a, b]$ has infinite dimension.

Verification of these statements is left to the reader.

**Exercises 1.1**

1. If $X$ is a linear space [Definition 1.1, (a)–(j)] show that for every $x \in X$

$$0x = 0$$
$$(-1)x = -x.$$

2. Show that the set of solutions of the differential equation

$$x'' + x = 0$$

is a linear space. What is the dimension of this space?

3. Show that $L_1$ and $l_1$ are linear spaces.

4. Show that the functions $1, e^t, e^{2t}, e^{3t}$ are linearly independent over any interval $[a, b]$.

5. Prove Theorem 1.1.

6. Show that $\dim(\mathscr{P}_n) = n + 1$.

7. Show that $C[a, b]$ has infinite dimension.

8. What is the dimension of the space spanned by

$$x_1 = \begin{pmatrix} 1 \\ 1 \\ 0 \end{pmatrix}, \quad x_2 = \begin{pmatrix} 0 \\ 1 \\ -1 \end{pmatrix}, \quad x_3 = \begin{pmatrix} 1 \\ 0 \\ 1 \end{pmatrix}?$$

9. Does the set of functions of the form $x(t) = 1/(a + bt)$ constitute a linear space?

10. Show that if $\{x_n\} \in l_p$, then $\{x_n\} \in l_{p'}$ when $p' > p > 0$.

11. Give an example of a function which is in $L_1[0, 1]$ but not in $L_2[0, 1]$.

## 1.2 NORMS

When analyzing approximation methods we often need to compare solutions or to measure the difference between various answers. In the terminology of linear spaces we must find the distance between two points in the space. Thus we want to generalize the notions of distance and vector length. The introduction of a *norm* provides this generalization.

**DEFINITION 1.5.** With each $x$ in a linear space $X$ we associate a nonnegative number, called the norm of $x$ and denoted by $\|x\|$, such that

   (a)  $\|x\| \geq 0$ for all $x \in X$.
   (b)  $\|x\| = 0$ if and only if $x = 0$.
   (c)  $\|\alpha x\| = |\alpha| \|x\|$ for all $x \in X$, $\alpha \in F$.
   (d)  The triangle inequality

$$\|x + y\| \leq \|x\| + \|y\|$$

   is satisfied for all elements $x, y \in X$.

Then $X$ is said to be a *normed linear space*. The nonnegative number $\|x - y\|$ is called the *distance* between the points $x$ and $y$.

**Example 1.3.** In $R^n$ we can define the norm of a vector

$$\mathbf{x} = \begin{bmatrix} x_1 \\ x_2 \\ \vdots \\ x_n \end{bmatrix}$$

by

$$\|\mathbf{x}\| = \left( \sum_{i=1}^{n} x_i^2 \right)^{1/2}.$$

We call this the Euclidean norm and it is the usual geometric length of the vector. More generally, for any integer $p \geq 1$ we can define a norm by

$$\|\mathbf{x}\|_p = \left( \sum_{i=1}^{n} |x_i|^p \right)^{1/p}. \tag{1.1}$$

To show that this does in fact define a norm we must show that conditions (a)–(d) of Definition 1.5 are satisfied. The first three of these are trivial; the triangle inequality (d) is essentially the well-known Minkowski inequality (Davis, 1963, p. 132)

$$\left( \sum_{i=1}^{n} |x_i + y_i|^p \right)^{1/p} \leq \left( \sum_{i=1}^{n} |x_i|^p \right)^{1/p} + \left( \sum_{i=1}^{n} |y_i|^p \right)^{1/p}. \tag{1.2}$$

The space $R^n$ with $p$-norm is sometimes denoted by $R_p^n$. $R_2^n$ is the familiar Euclidean $n$-space, but several other choices for $p$ are of importance in numerical analysis. With $p = 1$ we have

$$\|\mathbf{x}\|_1 = \sum_{i=1}^{n} |x_i|. \tag{1.3}$$

Here the length of a vector is measured by the sum of the magnitudes of its components. Another possibility is to use the magnitude of the largest component as a measure of length. Then

$$\|\mathbf{x}\|_\infty = \max_{1 \le i \le n} |x_i|. \tag{1.4}$$

This is the so-called maximum or infinity norm. The latter name, together with the notation, comes from the fact that

$$\|\mathbf{x}\|_p \to \|\mathbf{x}\|_\infty \qquad \text{as } p \to \infty.$$

(See Exercise 3 at the end of this section.)

**Example 1.4.** Analogous definitions can be made for the space $L_p$. For $x(t) \in L_p[a, b]$ we define the norm by

$$\|x\|_p = \left( \int_a^b |x(t)|^p \, dt \right)^{1/p}. \tag{1.5}$$

If $x(t)$ is bounded in $[a, b]$, the infinity norm is defined as

$$\|x\|_\infty = \sup_{a \le t \le b} |x(t)|. \tag{1.6}$$

In the language of abstract mathematics, by introducing a norm we have introduced a topology into the linear space and we can now generalize the geometric concepts such as neighborhoods, convergence, and so on. There are, of course, many other ways to define a topology. We could, for instance, use the more general notion of a metric to define the distance between two points, and thus be led to consider metric spaces instead of the less general normed spaces (see Exercise 5). However, there is little to be gained for our purpose in such a generalization; most of the problems in numerical analysis can be discussed adequately in the setting of normed linear spaces which we have chosen here.

We can now extend the idea of a basis to infinite-dimensional spaces.

**DEFINITION 1.6.** A set of elements $x_1, x_2, \ldots$ of a normed linear space $X$ is said to be *closed* (or complete) in $X$ if every $x \in X$ can be approximated arbitrarily closely (in norm) by a finite linear combination of the $x_i$. In other words, for any $x \in X$ and $\epsilon > 0$ there exists an $n$ and scalars $\alpha_1, \alpha_2, \ldots, \alpha_n$ such that

$$\left\| x - \sum_{i=1}^n \alpha_i x_i \right\| \le \epsilon.$$

If the set $\{x_i\}$ is closed in $X$ and linearly independent (that is, all finite subsets are linearly independent), then $\{x_i\}$ is called a basis for $X$ and we write

$$x = \sum_{i=1}^\infty \alpha_i x_i.$$

Spaces which possess a countable (or finite) basis are said to be *separable*. Spaces which are not separable play no significant role in numerical analysis and are therefore of no interest to us here. We shall always assume, without further mention, that all our spaces are separable.

**Exercises 1.2**

1. Determine which of the following expressions are valid as definitions for norms in $C^{(n)}[a, b]$.
   (a) $\max |x(t)| + \max |x'(t)|$.
   (b) $\max |x'(t)|$.
   (c) $|x(a)| + \max |x'(t)|$.
   (d) $|x(a)| \max |x(t)|$.
   (e) $\max_{a \le t \le b} \max_{0 \le r \le n} |x^{(r)}(t)|$.
2. In a normed linear space $X$ the set of all $x \in X$ such that

$$\|x\| = 1$$

   is called the unit ball (centered at the origin).
   (a) Sketch the unit balls in $R_1^2, R_2^2, R_{10}^2, R_\infty^2$.
   (b) Sketch the unit ball in the space $C[0, 1]$ with maximum norm. Can you draw a picture of the unit ball in $L_2[0, 1]$?
3. Justify the notation for the infinity norm by showing that for all $x \in R^n$

$$\lim_{p \to \infty} \|x\|_p = \max_{1 \le i \le n} |x_i|.$$

Do the same for $l_p$.

4. Two norms, $\| \ \|_a$ and $\| \ \|_b$, are said to be equivalent if there exist two positive constants $c_1$ and $c_2$, independent of $x$, such that

$$c_1 \|x\|_a \leq \|x\|_b \leq c_2 \|x\|_a.$$

Show that in $R^n$ the Euclidean norm and the maximum norm are equivalent. Can a similar claim be made for the norms defined by (1.5) and (1.6)?

5. A linear space $X$ is said to be a metric space if with every two elements $x,y \in X$ is associated a positive number $d(x,y)$ such that

$$d(x,y) \geq 0,$$
$$d(x,y) = 0 \qquad \text{if and only if } x = y,$$
$$d(x,y) = d(y,x),$$
$$d(x,y) \leq d(x,z) + d(z,y).$$

Every normed space is also a metric space if we use $\|x-y\|$ as the metric $d(x,y)$. Show by example that not every metric is a norm.

6. Show that

$$|\|x\| - \|y\|| \leq \|x - y\|.$$

7. A function from a linear space into $R^1$ satisfying (c) and (d) of Definition 1.5 is called a seminorm. Show that a seminorm must also satisfy (a) of Definition 1.5. Give an example of a seminorm which is not a norm.

8. Which of the expressions in Exercise 1 are seminorms, but not norms?

9. A norm $\| \ \|_a$ is said to be *stronger* than another norm $\| \ \|_b$ if $\lim_{n\to\infty}\|x_n\|_a = 0$ implies $\lim_{n\to\infty}\|x_n\|_b = 0$, but not vice versa. For $C[0,1]$ show that the maximum norm is stronger than the 2-norm.

## 1.3  BANACH AND HILBERT SPACES

The concept of a normed linear space is quite general, as can be seen from the various examples. In fact, in many cases it is too general to allow simple manipulations and it becomes appropriate to further restrict the types of spaces to be considered. We do this by requiring that our spaces possess some additional properties which will now be defined.

**DEFINITION 1.7.** A sequence $\{x_n\}$ in a normed linear space is said to be a *Cauchy sequence* if

$$\lim_{n\to\infty} \lim_{p\to\infty} \|x_{n+p} - x_n\| = 0. \tag{1.7}$$

**DEFINITION 1.8.** A normed linear space $X$ is said to be *complete* if every Cauchy sequence in $X$ converges to an element in $X$. A complete normed space is called a *Banach* space.

**Example 1.5.** The space $C[a,b]$ with the maximum norm is a Banach space. To show this let $\{x_n\}$ be a Cauchy sequence in $C[a,b]$. Then for sufficiently large $n$ and $p$

$$|x_{n+p}(t) - x_n(t)| \le \epsilon \qquad \text{for all } t \in [a,b].$$

From a standard result in advanced calculus (e.g., Rudin, 1964, p. 150) it then follows that the sequence converges uniformly to a limit which is continuous.

**Example 1.6.** The space $C[a,b]$ with the norm

$$\|x\| = \left( \int_a^b x^2(t)\, dt \right)^{1/2}$$

is not complete. For example, in $[-1, 1]$ the sequence

$$x_n(t) = \begin{cases} -1, & -1 \le t \le \dfrac{-1}{n} \\[2mm] nt, & \dfrac{-1}{n} < t < \dfrac{1}{n} \\[2mm] 1, & \dfrac{1}{n} \le t \le 1 \end{cases}$$

is a Cauchy sequence (Exercise 2), but the limit is clearly not continuous.

A complete space is obtained by admitting all Lebesgue-integrable functions, so that $L_2$, with the above norm, is a Banach space. This is a well-known result from analysis; a general proof showing that $L_p$, $1 \le p < \infty$ is a Banach space can be found in most texts on analysis (see for instance, Rudin, 1966). It is essentially this completeness property that makes it necessary to interpret the integrals as Lebesgue integrals, since the set of all Riemann-integrable functions is not complete in the $p$-norm.

**DEFINITION 1.9.** Let $X$ be a real linear space. With each $x, y \in X$ we associate a real number, denoted by $(x, y)$ and called the *inner product* of $x$ and $y$, such that

(a) $(x, x) \ge 0$ and $(x, x) = 0 \Rightarrow x = 0$.

(b) $(x, y) = (y, x)$.

(c) $(\alpha x, y) = \alpha(x, y)$.

(d) $(x + y, z) = (x, z) + (y, z)$.

**DEFINITION 1.10.**   A linear space on which an inner product is defined is said to be an *inner product space* (i.p.s.). A complete inner product space is referred to as a *Hilbert* space.

In an i.p.s. we always define the norm by

$$\|x\| = (x,x)^{1/2} \tag{1.8}$$

so that every i.p.s. is also a normed space. But the converse is not true; not every normed space can be thought of as an i.p.s. (Exercise 4).

With the introduction of a norm we were able to generalize the notion of distance between two points in a space. In an i.p.s. one can do considerably more and extend other geometric concepts, such as angles, orthogonality, and so on.

**THEOREM 1.2.**   For all $x,y$ in an i.p.s.

$$|(x,y)| \le \sqrt{(x,x)(y,y)} = \|x\| \|y\|. \tag{1.9}$$

This is the important *Schwarz inequality*.

PROOF.   Assume that $y \ne 0$, otherwise (1.9) is satisfied trivially. Then, by the definition of the inner product, for any $x,y,\lambda$,

$$(x+\lambda y, x+\lambda y) \ge 0$$

and

$$(x,x) + 2\lambda(x,y) + \lambda^2(y,y) \ge 0.$$

Choosing $\lambda = -(x,y)/(y,y)$ we get

$$(x,x) - \frac{(x,y)^2}{(y,y)} \ge 0$$

and (1.9) follows. Furthermore, if

$$(x,y)^2 = (x,x)(y,y),$$

then $x = \lambda y$. This follows by simply tracing the above steps backwards to show that $(x+\lambda y, x+\lambda y) = 0$.  ■

**DEFINITION 1.11.** The angle $\theta$ between two elements $x,y$ in an i.p.s. is defined by

$$\theta = \cos^{-1}\left[\frac{(x,y)}{\|x\|\,\|y\|}\right]. \tag{1.10}$$

This definition makes sense since, by the Schwarz inequality,

$$-1 \leq \frac{(x,y)}{\|x\|\,\|y\|} \leq 1.$$

**DEFINITION 1.12.** Two elements $x,y$ in an i.p.s. are said to be *orthogonal* if

$$(x,y) = 0$$

or, in geometrical terms, $\theta = \pi/2$. An element $x$ is said to be orthogonal to a subspace $\Phi$ if

$$(x,\varphi) = 0 \qquad \text{for all } \varphi \in \Phi.$$

**DEFINITION 1.13.** The element

$$\frac{(x,y)}{(x,x)}\,x$$

is the *orthogonal projection* of $y$ onto $x$.

**THEOREM 1.3.** *The Parallelogram Law.* For all $x,y$ in an i.p.s.

$$\|x+y\|^2 + \|x-y\|^2 = 2\|x\|^2 + 2\|y\|^2. \tag{1.11}$$

**THEOREM 1.4.** If

$$\|x+y\| = \|x\| + \|y\|, \tag{1.12}$$

then $x$ and $y$ are linearly dependent. Furthermore, if (1.12) is satisfied and $\|x\| = \|y\|$, then $x = y$.

The proofs of Theorems 1.3 and 1.4 are left as exercises.

**Example 1.7.** For **x** and **y** in $R^n$ we can define the inner product by

$$(\mathbf{x}, \mathbf{y}) = \sum_{i=1}^{n} x_i y_i. \tag{1.13}$$

The corresponding norm is then the Euclidean norm, and the definitions of angle and orthogonality have their usual geometric meaning.

A more general inner product, the *weighted* inner product, can be defined in $R^n$ by

$$(\mathbf{x}, \mathbf{y}) = \sum_{i=1}^{n} w_i x_i y_i. \tag{1.14}$$

For $w_i > 0$ this defines a proper inner product.

**Example 1.8.** For $L_2[a,b]$ we can define the inner product by

$$(x,y) = \int_a^b x(t) y(t) \, dt. \tag{1.15}$$

Since the space is complete under the 2-norm, $L_2[a,b]$ is a Hilbert space.

**Exercises 1.3.**

1. Prove that every finite-dimensional normed linear space is complete.
2. Show that the sequence $\{x_n\}$ in Example 1.6 is a Cauchy sequence.
3. Show that if $\{x_n\}$ is a Cauchy's sequence, then for all sufficiently large $n$, $\|x_n\| \le K < \infty$.
4. Show that in $C[a,b]$ with maximum norm we cannot define an inner product such that

$$(x,x)^{1/2} = \|x\|_\infty$$

for all $x$.
5. Prove Theorems 1.3 and 1.4.
6. In $C^{(1)}[a,b]$ is

$$(x,y) = \int_a^b x'(t) y'(t) \, dt + x(a) y(a)$$

an inner product?

7. Show that in $C[a,b]$

$$(x,y) = \int_a^b w(t)x(t)y(t)\,dt$$

with $w(t)>0$ defines an inner product. Is this also true if $w(t)$ is not strictly positive in $[a,b]$?

8. Does the parallelogram law hold in the space $C[a,b]$ with maximum norm?

9. Let $z$ be the orthogonal projection of $x$ onto $y$. Show that $(x-z)$ is orthogonal to $y$.

10. Assume that $\varphi_1, \varphi_2, \ldots$ is a basis for an i.p.s. $X$. If

$$(x,\varphi_i) = 0, \qquad i = 1,2,\ldots$$

show that $x=0$.

11. Which of the following are Banach spaces?
    (a) $C^{(1)}[0,1]$ with maximum norm.
    (b) $C^{(1)}[0,1]$ with norm $\|x\| = \|x\|_\infty + \|x'\|_\infty$.
    (c) $C^{(1)}[0,1]$ with norm $\|x\| = |x(0)| + \|x'\|_\infty$.

## 1.4 MAPPINGS AND OPERATORS

Let $S_1$ be a subset of some linear space $X$ and let $S_2$ be a subset of another (or the same) linear space $Y$. Assume that we have a rule such that with each element of $S_1$ there is associated a unique element of $S_2$. Then such a rule is called a *mapping* or a *transformation* from $S_1$ to $S_2$. The representation of such a transformation will be called an *operator*. We will use capital letters to denote operators; the operator equation

$$Tx = y$$

denotes that the operator $T$ maps the element $x \in S_1$ to an element $y \in S_2$. Occasionally we want to represent the operator without explicitly showing the element on which it acts; in that case we use the "place-holder" notation, where [ ] represents the place where the element on which the operator acts is to be inserted. Thus $T[\ ]$ would represent the above operator.

The set of all elements for which an operator $T$ is defined is its *domain* $\mathcal{D}(T)$. The set of all elements generated by $T$, that is, the set of all elements $y$ such that $y = Tx$ for some $x \in \mathcal{D}(T)$, is the *range* of $T$, denoted by $\mathcal{R}(T)$.

To show that $T$ is a mapping from $S_1$ to $S_2$ we will use the notation $T : S_1 \rightarrow S_2$.

If $S_2$ contains no elements except those in $\mathcal{R}(T)$, then we say that the mapping $T$ is *onto* $S_2$; a mapping which is not necessarily onto is said to be *into*.

**Example 1.9.**

(a) An $n \times m$ matrix with the usual definition of vector-matrix multiplication is an operator from $R^m$ into $R^n$. Whether the mapping is onto $R^n$ depends on the rank of the matrix.

(b) $d/dt[\ ]$ is an operator from the space of continuously differentiable functions into the space of continuous functions.

(c) The operator $\int_0^1 [\ ]\,dt$ maps the space of integrable functions into $R^1$.

**DEFINITION 1.14.**   An operator $L$ is said to be *linear* if its domain is a linear space and if

$$L(\alpha x + \beta y) = \alpha L x + \beta L y \qquad (1.16)$$

for all $x, y \in \mathcal{D}(L)$ and all scalars $\alpha, \beta$.

Linearity is an important property; in classical applied mathematics a considerable body of results has been developed for linear problems. The nonlinear case, on the other hand, poses a great many theoretical and practical difficulties. Not surprisingly, the same state of affairs exists in numerical analysis, as we shall see.

**DEFINITION 1.15.**   An operator $T$ is said to be *one-to-one* if

$$T x_1 = T x_2 \Rightarrow x_1 = x_2. \qquad (1.17)$$

Thus if an operator is one-to-one, each distinct element is $\mathcal{D}(T)$ maps into a distinct element in $\mathcal{R}(T)$. In other words, to each element in $\mathcal{R}(T)$ there corresponds a unique element in $\mathcal{D}(T)$, and we can think of this correspondence as a mapping from $\mathcal{R}(T)$ into $\mathcal{D}(T)$. The operator representing this mapping is called the *inverse* of $T$ and is denoted by $T^{-1}$. We then have the following result.

**THEOREM 1.5.**   Let $T : X \rightarrow Y$ be one-to-one and onto $Y$. Then $T$ has an inverse whose domain includes $Y$.

We can now define addition and scalar multiplication of operators. If $T_1$ and $T_2$ are two operators with a common domain $X$, then $(T_1 + T_2)$ is the operator such that

$$(T_1 + T_2)x = T_1 x + T_2 x \qquad \text{for all } x \in X. \qquad (1.18)$$

Similarly, $(\alpha T)$ is defined by

$$(\alpha T)x = \alpha(Tx). \qquad (1.19)$$

Under certain circumstances one can also define the product of two operators. Let $R$ and $T$ be two operators such that $\mathcal{R}(T)$ is contained in $\mathcal{D}(R)$. Then $RT$ is defined by

$$(RT)x = R(Tx). \qquad (1.20)$$

It should be noted that operators do not always satisfy the commutative law, that is $RT \neq TR$ generally. In fact $TR$ may not exist even though $RT$ does. Matrix theory provides the most familiar example of this observation.

If an operator is invertible, then it is clear from the definition of the inverse that

$$T^{-1}T = I \qquad (1.21)$$

and

$$TT^{-1} = I \qquad (1.22)$$

where $I$ is the *identity* operator defined by

$$Ix = x \qquad \text{for all } x. \qquad (1.23)$$

For operators that do not have an inverse it may be possible to introduce a less restrictive concept.

**DEFINITION 1.16.** If there exists an operator $T_L^{-1}$ such that

$$T_L^{-1}T = I, \qquad (1.24)$$

then $T_L^{-1}$ is said to be a *left inverse* of $T$. If there exists an operator $T_R^{-1}$ such that

$$TT_R^{-1} = I, \qquad (1.25)$$

then $T_R^{-1}$ is said to be a *right inverse* of $T$.

The following properties of inverses are easily proved (Exercise 12).

**THEOREM 1.6**

(a)  If $T_L^{-1}$ and $T_R^{-1}$ both exist, then $T_R^{-1} = T_L^{-1} = T^{-1}$.
(b)  If $T_R^{-1}$ exists, then $T$ is onto $Y$ and the equation $Tx = y$ has at least one solution $x = T_R^{-1}y$.
(c)  If $T_L^{-1}$ exists, then $T$ is one-to-one. If the equation $Tx = y$ has a solution, then that solution is unique and is given by $x = T_L^{-1}y$.

**DEFINITION 1.17.** An operator $T : X \rightarrow Y$ is said to be *bounded* if and only if there exists a $c < \infty$ such that

$$\| Tx_1 - Tx_2 \| \leq c \| x_1 - x_2 \| \tag{1.26}$$

for all $x_1, x_2$ in $X$. The value $\mu(T) = \inf c$ is called the *bound* of the operator $T$ on $X$.

An operator for which (1.26) holds only if $x_1$ and $x_2$ are in some subset $S$ of $X$ is said to be *bounded on S*.

For linear operators the definition and the computation of the bound can be simplified.

**THEOREM 1.7.**  Let $L$ be a linear operator from $X$ into $Y$. Then

$$\mu(L) = \sup_{x \in X, x \neq 0} \frac{\| Lx \|}{\| x \|} = \sup_{\| x \| = 1} \| Lx \|. \tag{1.27}$$

**PROOF.**  Since $L$ is linear we have from (1.26)

$$\| Lx \| \leq \mu(L) \| x \| \qquad \text{for all } x \in X$$

where $\mu(L)$ is the smallest possible value. Let

$$\mu' = \sup_{x \in X, x \neq 0} \frac{\| Lx \|}{\| x \|},$$

then $\| Lx \| \leq \mu' \| x \|$ and there exists no $\mu < \mu'$ such that this inequality is satisfied for all $x$. Hence

$$\mu' = \mu(L).$$

Also, we can write $x = x\|x\|/\|x\|$, so that

$$\frac{\|Lx\|}{\|x\|} = \frac{\|x\| \|L(x/\|x\|)\|}{\|x\| \|x/\|x\|\|}$$

which proves the second part of (1.27). ∎

The set of all bounded linear operators from a linear space $X$ into a linear space $Y$, with addition and scalar multiplication defined by (1.18) and (1.19), is itself a linear space (Exercise 3). This space is generally denoted by $L[X, Y]$. $L[X, X]$ is frequently abbreviated as $L[X]$ or just $[X]$.

For linear operators it is customary to talk about the norm of the operator instead of a bound.

**DEFINITION 1.18.** Let $L$ be a linear operator on $X$. Then the norm of $L$, denoted by $\|L\|$, is defined by

$$\|L\| = \sup_{\|x\| = 1} \|Lx\|. \tag{1.28}$$

This implies that

$$\|Lx\| \leq \|L\| \|x\| \tag{1.29}$$

and that there exists an $x$ such that equality either is attained or is approached arbitrarily closely. The operator norm depends on the norm in $X$ and is said to be *induced* by that norm.

**Example 1.10.** Let $A$ be an $n \times n$ matrix with elements $a_{ij}$. For the vector norm

$$\|\mathbf{x}\| = \max_i |x_i|$$

the induced matrix norm is

$$\|A\| = \max_i \sum_{j=1}^{n} |a_{ij}|. \tag{1.30}$$

PROOF.

$$\|A\mathbf{x}\| = \max_i \left| \sum_{j=1}^{n} a_{ij} x_j \right| \leq \max_i \sum_{j=1}^{n} |a_{ij}| |x_j| \leq \|\mathbf{x}\| \max_i \sum_{j=1}^{n} |a_{ij}|.$$

To complete the proof we show that there exists an $x$ such that equality is attained. Suppose $\sum_{j=1}^{n} |a_{ij}|$ takes on its maximum for $i = k$. Take $x$ to be the vector with components

$$x_j = 1 \qquad \text{if } a_{kj} \geq 0$$
$$x_j = -1 \qquad \text{if } a_{kj} < 0.$$

Then

$$\sum_{j=1}^{n} a_{kj} x_j = \sum_{j=1}^{n} |a_{kj}|$$

and

$$\|A\mathbf{x}\| = \sum_{j=1}^{n} |a_{kj}| = \|\mathbf{x}\| \sum_{j=1}^{n} |a_{kj}|. \qquad \blacksquare$$

**Example 1.11.** Consider the integral operator $\mathcal{K} : C[0,1] \to C[0,1]$ defined by

$$(\mathcal{K}x)(t) = \int_0^1 K(t,s)x(s)\,ds,$$

where $K(t,s)$ is a continuous function of $t$ and $s$. Using the maximum norm for $x$,

$$\|x\|_\infty = \max_{0 \leq t \leq 1} |x(t)|,$$

the induced operator norm is

$$\|\mathcal{K}\|_\infty = \max_{0 \leq t \leq 1} \int_0^1 |K(t,s)|\,ds. \qquad (1.31)$$

The proof is left as an exercise.

**DEFINITION 1.19.** Let $\{x_n\}$ be a sequence of elements in a linear space $X$. We say this sequence converges to $x$, denoted by $x_n \to x$, if and only if

$$\lim_{n \to \infty} \|x - x_n\| = 0.$$

**DEFINITION 1.20.** We say that an operator $T$ is *continuous* at $x$ if for every sequence $\{x_n\}$ converging to $x$

$$\lim_{n \to \infty} \|Tx_n - Tx\| = 0.$$

**THEOREM 1.8.** A linear operator $L$ is continuous everywhere if and only if it is bounded.

PROOF. Assume that $L$ is bounded. Then

$$\|Lx - Lx_n\| \leq \|L\|\,\|x - x_n\|$$

so that $\|Lx_n - Lx\| \to 0$ if $x_n \to x$, and hence $L$ is continuous everywhere. Now suppose that $L$ is continuous at some point $x$. Then, since

$$u_n = (x - x_n) \to 0 \Rightarrow \|Lu_n\| \to 0,$$

$L$ is also continuous at 0. Thus we can find a $\delta > 0$ such that

$$\|Lz\| \leq 1 \qquad \text{for all } \|z\| \leq \delta.$$

Hence, for any $x \neq 0$,

$$\|Lx\| = \frac{\|x\|}{\delta}\left\|L\frac{\delta x}{\|x\|}\right\| \leq \frac{\|x\|}{\delta}$$

so that $L$ is bounded and $\|L\| \leq 1/\delta$. It also follows that a linear operator which is continuous at one point is continuous everywhere. ∎

For linear operators the concepts of continuity and boundedness are identical, but this is not true for nonlinear operators. It follows from Definition 1.17 that every bounded operator is also continuous, but one can easily find examples to show that the converse is not true. It is instructive to consider what these two concepts mean when the operator is a function of one variable. The definition of continuity is the usual one, but boundedness implies that

$$|f(t_1) - f(t_2)| \leq c|t_1 - t_2|,$$

which is generally known as Lipschitz continuity and is a stronger condition than ordinary continuity.

It should be pointed out that we have taken a few liberties with the notation. For instance, we have used the same symbol for the norm in $X$ and $Y$ although these norms may be defined quite differently. Of course, the context always clarifies the meaning and we may, for the sake of simplicity, ignore such fine notational distinctions.

**Exercises 1.4.**

1. Show that the operator $d/dx[\quad]:C^{(1)}[a,b]\to C[a,b]$ is onto $C[a,b]$ but not one-to-one. What is a right inverse of this operator?

2. What is the relation between the rank of a matrix $A$ and the question whether $A:R^m\to R^n$ is onto or one-to-one?

3. Show that $L[X]$ is a linear space.

4. Show that the operator norm, defined by (1.28), does indeed satisfy all the requirements of a norm.

5. For a matrix $A:R^n\to R^n$ show that the vector norm $\|x\|_1$ induces the matrix norm

$$\|A\| = \max_j \sum_{i=1}^n |a_{ij}|.$$

6. (a) For $\mathcal{K}:C[0,1]\to C[0,1]$ prove Equation 1.31, Example 1.11.
   (b) If the 1-norm is used for $x$, show that

$$\|\mathcal{K}\| = \max_{0\le s\le 1} \int_0^1 |K(t,s)|\,dt.$$

7. If $L$ is a linear operator show that $L(0)=0$.

8. For the operator $D[\quad]=d/dt[\quad]:C^{(1)}[a,b]\to C[a,b]$
   (a) Show that if we use the infinity norm in $C^{(1)}$ and $C$, then $D$ is not a bounded operator.
   (b) Define appropriate norms in $C^{(1)}$ and $C$ such that $D$ is bounded and compute $\|D\|$.

9. Find an example of a nonlinear operator $T:R^2\to R^2$ which is continuous at $(0,0)$ but not bounded in any neighborhood of the origin.

10. Find a sequence of functions $f_n(t)$ for which

$$\|f_n\|_1 \to 0$$
$$\|f_n\|_\infty \to \infty, \qquad \text{as } n\to\infty.$$

11. For $A,B\in L[X]$, show that

$$\|AB\| \le \|A\|\,\|B\|$$
$$\|A^n\| \le \|A\|^n, \qquad \text{for all } n\ge 1.$$

12. Prove Theorem 1.6.

13. (a)  If $L$ is a linear operator, show that $\mathcal{R}(L)$ is a linear space.
    (b)  Show, by example, that this is not necessarily true for nonlinear operators.

## 1.5 CLASSIFICATION OF PROBLEMS IN COMPUTATIONAL MATHEMATICS

We can now make a rather broad, but still useful classification of the types of problems one encounters in computational mathematics. A large number of problems arising in applied mathematics and therefore of interest to the numerical analyst can be stated formally as: Solve the equation

$$Tx = y \qquad (1.32)$$

where $x \in X, y \in Y$, $X$ and $Y$ are linear spaces, and $T: X \to Y$. To elaborate, we can recognize three distinct types of problems.

1. *The direct problem.* Given $T$ and $x$, find $y$. The computation of the definite integral is an example.
2. *The inverse problem.* Given $T$ and $y$, find $x$. Solving systems of simultaneous equations, ordinary and partial differential equations, and integral equations are examples.
3. *The identification problem.* Given $x$ and $y$, find $T$.

In the language of systems engineering, $x$, $y$, and $T$ represent the input, output, and the system, respectively. Thus, in a direct problem, we are trying to determine the output of a given system generated by a known input; in the inverse problem one looks for the input which generates a known output. In the identification problem one tries to find the laws governing a system from a knowledge of the relation between input and output (generally we know only a finite number of input-output pairs).

Each type of problem generates its own specific questions, and classes of algorithms and success in developing a general theory has varied.

Direct problems are relatively easily treated. Numerical integration is the main example and this is now quite well understood. This is not to say that all difficulties have been resolved. There are still many theoretical and practical questions unanswered, as for instance in the evaluation of $n$-dimensional integrals with $n$ fairly large. Nevertheless, the problem is conceptually simple and a general theory not very complicated as we will see in Chapter 3.

The inverse problem, because of its importance in applications, occupies a central place in numerical analysis. The linear case, in particular, has been studied extensively and its theory is well-developed. The situation is somewhat less satisfactory in the nonlinear case. Chapters 4 and 5 will deal with the inverse problem.

The identification problem in a general setting is rather difficult. This is not surprising when we consider its interpretation: from a finite number of observations we are trying to infer the laws governing an unknown "black-box" system. This is generally impossible unless one has specific information on the structure of the system. The identification problem in its simplest form is a topic in approximation theory, which will be discussed in Chapter 2. Here the operator is a function of one variable and we are asked to find the function whose graph passes through (or close to) a set of points $\{x_i, y_i\}$. Clearly, this question cannot be answered uniquely unless we restrict the class of admissible functions. In the more complicated case where $X$ and $Y$ are infinite-dimensional spaces, the best one can do is to assume some plausible form for $T$, include in it some undetermined parameters, and then compute the values of these parameters which best explain the observed input-output relationship. This leads to the topics of minimization of functionals, mathematical programming, and statistics. Except for a brief discussion of the fundamentals of minimization in Chapter 5 we will not pursue this topic.

In certain simple examples the inverse problem $Tx = y$ can be converted into a direct problem. If $T$ has a known inverse then $x = T^{-1}y$. For example, the solution of the differential equation

$$\frac{dx}{dt} = f(t)x, \qquad x(0) = 1,$$

can be found, by separation of variables, to be

$$x(t) = \exp\left( \int_0^t f(s)\, ds \right).$$

At this point the traditional mathematician might consider the problem solved, but unless $f(s)$ is simple enough to have a known indefinite integral the original differential equation is probably more convenient for numerical computation. Another, less trivial, example is the representation of the solution of certain elliptic partial differential equations via Green's functions which, while theoretically interesting, may be inconvenient numerically. This is not to say that the inverted form is useless, often it provides insight into the qualitative behavior of the solution which cannot be obtained otherwise. Still, from a strictly numerical point of view, the explicit determination of the inverse does not necessarily help.

At first sight the formulation (1.32) may seem too restrictive, since often one has not only an equation to solve, but also initial or boundary conditions to satisfy. To allow us to represent more complicated cases in our simple notation we introduce the idea of a product space.

**DEFINITION 1.21.** Let $X$ and $Y$ be two linear spaces. Then their *product space*, denoted by $X \times Y$, is the set of all ordered pairs $(x,y)$, with $x \in X$, $y \in Y$.

Addition and scalar multiplication in a product space are defined in a natural way. If $z_1 = (x_1, y_1)$ and $z_2 = (x_2, y_2)$, then $z_1 + z_2 = (x_1 + x_2, y_1 + y_2)$ and $\alpha z_1 = (\alpha x_1, \alpha y_1)$. It is then easy to show that $X \times Y$ satisfies all the postulates of a linear space. Using the idea of a product space initial and boundary conditions can then be included in the general form (1.32) as the following example shows.

**Example 1.12.** Consider the differential equation

$$\frac{dx(t)}{dt} = f(t, x(t)), \qquad 0 \le t \le 1,$$

with initial condition $x(0) = \alpha$.

Let $Y = C[0, 1] \times R^1$. Then we define $T: C^{(1)}[0, 1] \to Y$ by

$$Tx = \left( \frac{dx}{dt} - f(t, x), x(0) \right)$$

and the equation can be written as

$$Tx = (0, \alpha).$$

When the spaces involved in (1.32) are infinite-dimensional a direct numerical treatment is generally not possible. The problem must be recast in the more suitable form

$$T_n x_n = y_n \tag{1.33}$$

where $x_n \in X_n$, $y_n \in Y_n$, and $X_n$ and $Y_n$ are finite-dimensional spaces. This process is called *discretization*. Since (1.32) and (1.33) are not equivalent, discretization introduces an error. Furthermore, the discretized problem usually cannot be solved exactly, and further error arises from the limitations of the computational procedure. These two sources combine to cause the overall error in the final result.

*Computational errors* are due to the fact that the process required to solve (1.33) cannot be carried out exactly. The fact that arithmetic operations can be performed only with limited precision introduces the problem of *round-off error*. While it is easy to discuss the behavior of round-off error in each particular operation, its cumulative effect on a lengthy sequence of calculations is hard to predict. The most extensive analysis of round-off error is given in J. H. Wilkinson's classic treatment (Wilkinson, 1963), but a comprehensive and universally useful theory has not yet emerged. We will therefore not pursue this subject, and instead refer the interested reader to Wilkinson's work.

There may be other sources of computational error; for instance, the solution of (1.33) may call for an infinite iteration process, so that some error will be introduced by stopping after a finite number of steps. This is much easier to analyze, as will be seen from specific examples to be discussed.

*Discretization errors* arise because we replace an infinite or continuous process by a finite one. Discretization usually depends on one or more parameters, represented by $n$ in (1.33). For example, in the numerical evaluation of a definite integral (a continuous process) by numerical quadrature (a finite summation), $n$ might represent the number of quadrature points. In studying the discretization error we need to investigate the relation of the true solution (say $x$) of (1.32) to the approximate solution $x_n$ of (1.33). It is reasonable to require that a numerical method be capable of yielding, at least in the absence of computational errors, arbitrarily accurate answers, obtainable by making the discretization sufficiently fine (which we will denote by $n \to \infty$). In other words, we expect the approximate solution to converge to the true solution or

$$\lim_{n \to \infty} \|x - x_n\| = 0.$$

A method which gives a sequence of approximations converging to the true answer is called a *convergent approximation method*.

More detailed information about the error is usually needed in practical applications. A function $B(n)$ such that

$$\|x - x_n\| \le B(n)$$

is called an *error bound*. Often error bounds can be expressed in the form

$$\|x - x_n\| \le Cn^{-p} \qquad \text{for all } n > 0$$

where $C$ is independent of $n$. Clearly, if $p > 0$, the method is convergent. If

$p$ is the largest number for which such an inequality can hold, then we say that $p$ is the *order of convergence* of the method.

The actual form of the error bound is usually somewhat complicated, and often there are alternate ways in which it can be specified. Typically, we might have $B(n) = B(n, x)$, that is, the bound is expressed in terms of the solution of the original problem. This form tends to be useful in proving convergence, but it is not easy to obtain an actual numeric bound on the error from it. In some cases it is possible to get a bound of the form $B(n) = B(n, x_n)$. Now the bound is expressed in terms of the known approximate solution; hence, this is sometimes called an *a posteriori* error bound. *A posteriori* error bounds, together with some information or assumptions about the original problem, can often be used to obtain explicit, computable bounds on the error. This will be demonstrated with some examples in later sections.

Error bounds can, and often do, seriously overestimate the actual error. If this is the case, the bound is said to be *pessimistic*. A bound which closely estimates the actual error (within an order of magnitude, say) is called *realistic*.

If we can write

$$x - x_n = e_n + g_n$$

such that

$$\lim_{n \to \infty} \frac{\| g_n \|}{\| e_n \|} = 0,$$

then $e_n$ is said to be an *asymptotic error estimate*, since for large $n$, $e_n$ is a good approximation to the actual error.

Again, it should be pointed out that the notation adopted above lacks full generality. For example, $x$ and $x_n$ may not be in the same spaces, so that their difference is not necessarily defined. Similarly, expressions like $n^{-p}$ and $n \to \infty$ have to be properly interpreted when there are several discretization parameters involved. However, this approach is in keeping with the spirit of this section, which aims to show some of the overall structure of the problem without undue attention to detail. In the specific cases to be discussed later we shall need to be more careful.

The simple ideas introduced in this chapter provide us with sufficient basics to begin our study of the theoretical aspects of numerical analysis, but before doing so we will review some of the results from approximation theory. Approximation theory is of fundamental importance in numerical analysis, since in many of the problems the underlying spaces are function spaces and discretization involves approximating these with simpler, finite-dimensional spaces. Thus, a knowledge of the most important results of approximation theory is indispensable.

### Exercises 1.5

1. Write the following problem in general operator notation:

$$\frac{\partial^2 \varphi}{\partial t^2} + \frac{\partial^2 \varphi}{\partial s^2} = 0$$

with

$$\varphi(t,s) = ts \qquad \text{when } t^2 + s^2 = 1.$$

2. (a)  If $X$ and $Y$ are linear spaces, show that $X \times Y$ is also a linear space.
   (b)  If $\dim(X) = n$ and $\dim(Y) = m$, show that $\dim(X \times Y) = m + n$.
3. If $\| \ \|_X$ and $\| \ \|_Y$ denote norms in $X$ and $Y$, respectively, show that in $X \times Y$ we can define a norm by

$$\|z\| = \max(\|x\|_X, \|y\|_Y)$$

where $z = (x,y)$.
4. If two norms are equivalent show that convergence in one norm implies convergence in the other.
5. Review the nature of floating point arithmetic for some typical digital computer. Discuss which of the laws of algebra (associative, distributive, commutative) are not satisfied for floating point arithmetic.

# 2
# APPROXIMATION THEORY

The theory of the approximation of functions of one or more variables is one of the major topics of mathematics. It has received the attention of many of the great mathematicians of the nineteenth and twentieth centuries, and many elegant and far-reaching results have been developed in the last 150 years. Approximation theory has certainly proved to be a cornerstone of numerical analysis, essential to the development of most algorithms. In our study we are not able to delve deeply into the theory, but we try to single out those results that have been of greatest importance in computational mathematics. As we are mainly concerned with results, rather than further development, we often omit proofs, especially those that are highly technical and contribute little to the understanding of the final result. All of these proofs can of course be found in standard texts on approximation theory to which we refer.

Our aim then is to investigate the approximation of functions from certain infinite-dimensional spaces (primarily $C[a,b]$) by means of simpler functions, generally coming from a finite-dimensional space. From the standpoint of the numerical analyst the approximating space should have certain properties. It should be possible to get arbitrarily good approximations to a given function by making the dimension of the approximating space sufficiently large, that is, the approximations should converge in the sense of Section 1.5. Elements of the approximating space should be simple so that they can be easily manipulated (e.g., integrated and differentiated). Finally, there should exist a well-developed theory to facilitate the analysis of the resulting computational procedures. Polynomials are an ideal choice on all three counts. That they are quite powerful in approximating functions is demonstrated by the following classical result.

**THEOREM 2.1 WEIERSTRASS APPROXIMATION THEOREM.** Let $f(t) \in C[a,b]$. Then for any $\epsilon > 0$ there exists a polynomial $P_n(t)$ such that

$$\max_{a \le t \le b} |f(t) - P_n(t)| \le \epsilon.$$

There are a number of proofs of this important result; for one approach see Davis (1963, p. 107).

While the theorem is simply an existence theorem, some of the proofs, such as the one given by Davis, are constructive and provide an explicit representation of a sequence of polynomials converging to the given function. Unfortunately, the convergence is usually so slow that little practical use can be made of such representations and the question of the construction of the approximating polynomial must be considered in more detail.

Other types of approximations have been found useful. The classical Fourier theory which uses trigonometric functions as approximants is an important tool in applied mathematics. Recently, rational approximations, that is, approximation by functions which are ratios of polynomials, have begun to interest the numerical analyst. But none of these has managed to displace polynomials from their central role in the design of computational algorithms.

Perhaps the best known method of approximation is polynomial interpolation, which consists of finding a polynomial $P_n(t)$ taking on pre-assigned values $w_i$ at certain points $t_i$. We consider polynomial interpolation in detail in Section 2.2, but before doing so we discuss the interpolation problem from a more general point of view.

## 2.1   THE GENERAL INTERPOLATION PROBLEM

**DEFINITION 2.1.**   A mapping from a linear space $X$ into $R^1$ is called a *functional*.

Obviously a functional is just a special case of an operator, and all results and definitions of Chapter 1 hold. Thus the set of all bounded linear functionals on a linear space $X$ is itself a linear space, called the *algebraic conjugate or dual space on $X$* and denoted by $X^*$.

The general interpolation problem can then be stated as: Let $X$ be a space of dimension $n$, and let $L_1, L_2, \ldots, L_n$ be given linear functionals in $X^*$. Find an $x \in X$ such that

$$L_i x = w_i, \qquad i = 1, 2, \ldots, n, \tag{2.1}$$

where the $w_i$ are given.

**Example 2.1.**   The form (2.1) of the interpolation problem is a generalization of the classical polynomial interpolation mentioned above. To make the connection we take $\mathcal{P}_{n-1}$ as $X$ and define the functionals by

$$L_i x = x(t_i),$$

where the $t_i$ are a set of distinct points. The interpolation problem then is to find a polynomial $P_{n-1}(t)$ taking on preassigned values $w_1, w_2, \ldots, w_n$ at the points $t_1, t_2, \ldots, t_n$. This type of interpolation is called *Lagrange interpolation*.

The statement of the problem in its general form (2.1) immediately raises some questions. Is the problem solvable? Is the solution unique? If interpolation is used to approximate a function, how accurate is the answer? The first two of these questions will be discussed in a general setting in this section. The last question, however, is much more difficult and we are able to give useful answers only for certain types of polynomial interpolation.

The algebraic conjugate space $X^*$ is a linear space, so linear independence is defined in the usual way, that is, the functionals $L_1, L_2, \ldots, L_n$ are linearly independent if and only if

$$\alpha_1 L_1 + \alpha_2 L_2 + \cdots + \alpha_n L_n = 0 \Rightarrow \alpha_1 = \alpha_2 = \cdots = \alpha_n = 0.$$

The 0 in this linear combination of the functionals denotes the zero functional, that is, the functional mapping every $x \in X$ into 0.

**Example 2.2.** The functionals on $C^{(1)}[a,b]$ defined by

$$L_1 x = x(c), \qquad a \le c \le b$$
$$L_2 x = x'(c)$$
$$L_3 x = \int_a^b x(t)\, dt$$

are linearly independent. To show this, assume that

$$\alpha_1 L_1 + \alpha_2 L_2 + \alpha_3 L_3 = 0$$

then

$$(\alpha_1 L_1 + \alpha_2 L_2 + \alpha_3 L_3) x = 0, \qquad \text{for all } x \in C^{(1)}[a,b].$$

Now choose an $x$ such that $x(c) = x'(c) = 0$, and $x(t) \ge 0$. This implies that $\alpha_3 = 0$. By similar arguments we can show that $\alpha_1 = \alpha_2 = 0$, hence the functionals are linearly independent.

On the other hand, if we had used $\mathcal{P}_1$ as $X$, then the three functionals would have been linearly dependent. This can be shown directly (Exercise 3); it also follows from the following theorem.

**THEOREM 2.2.** If $\dim(X) = n < \infty$, then $\dim(X^*) = n$.

**PROOF.** See Davis (1963, p. 18). ∎

Thus, in an $n$-dimensional space any linear functional can be expressed as a linear combination of $n$ linearly independent functionals, or any $n$ linearly independent functionals are a basis for $X^*$.

**THEOREM 2.3.** Let $X$ be an $n$-dimensional space, and $x_1, x_2, \ldots, x_n \in X$ be linearly independent. Then $L_1, L_2, \ldots, L_n \in X^*$ are linearly independent if and only if

$$\det(L_i x_j) = \begin{vmatrix} L_1 x_1 & \cdots & L_1 x_n \\ \vdots & & \vdots \\ L_n x_1 & \cdots & L_n x_n \end{vmatrix} \neq 0.$$

**PROOF.** Let the $L_i$ be linearly independent and assume that $\det(L_i x_j) = 0$. Then $\det(L_j x_i) = 0$, which implies that the system

$$\begin{aligned} a_1 L_1 x_1 + a_2 L_2 x_1 + \cdots + a_n L_n x_1 &= 0 \\ \vdots & \qquad \vdots \\ a_1 L_1 x_n + a_2 L_2 x_n + \cdots + a_n L_n x_n &= 0 \end{aligned} \tag{2.2}$$

has a nontrivial solution $a_1, a_2, \ldots, a_n$ such that

$$(a_1 L_1 + a_2 L_2 + \cdots + a_n L_n) x_i = 0, \qquad \text{for } i = 1, 2, \ldots, n.$$

But the $x_i$ are linearly independent; hence, they are a basis for $X$ and any $x \in X$ can be expressed as a linear combination of the $x_i$. Thus, there exists a nontrivial set $a_i$ such that

$$(a_1 L_1 + a_2 L_2 + \cdots + a_n L_n) x = 0$$

for all $x \in X$, which contradicts the assumption that the $L_i$ are linearly independent. Hence we must have $\det(L_i x_j) \neq 0$.

Now assume that the $L_i$ are linearly dependent. Then there exists a nontrivial set $\{a_i\}$ such that

$$(a_1 L_1 + a_2 L_2 + \cdots + a_n L_n) x = 0$$

for all $x \in X$, particularly for $x_1, x_2, \ldots, x_n$. Thus, (2.2) has a nontrivial solution, implying $\det(L_i x_j) = 0$. So since

$$\{L_i\} \text{ linearly dependent} \Rightarrow \det(L_i x_j) = 0,$$

we have

$$\det(L_i x_j) \neq 0 \Rightarrow \{L_i\} \text{ linearly independent,}$$

and the proof is complete. ∎

**THEOREM 2.4.** Let $\dim(X) = n$. Then the interpolation problem (2.1) has a unique solution if and only if the $L_i$ are linearly independent in $X^*$.

PROOF. Let $x_1, x_2, \ldots, x_n$ be a basis for $X$ and consider the system

$$L_i(a_1 x_1 + a_2 x_2 + \cdots + a_n x_n) = w_i, \qquad i = 1, 2, \ldots, n.$$

If the $L_i$ are linearly independent, then $\det(L_i x_j) \neq 0$ by the previous theorem and hence the system has a unique solution. Conversely, if the system does have a unique solution, then $\det(L_i x_j) \neq 0$ and the $L_i$ are linearly independent. We call $\det(L_i x_j)$ the *generalized Gram determinant*. ∎

A method for solving the interpolation problem is contained in the above proof. Let $x_1, x_2, \ldots, x_n$ be a basis for $X$. Then the solution to (2.1) is given by

$$x = a_1 x_1 + a_2 x_2 + \cdots + a_n x_n,$$

with the coefficient $a_i$ determined from the linear system

$$(L_i x_1) a_1 + (L_i x_2) a_2 + \cdots + (L_i x_n) a_n = w_i, \qquad i = 1, 2, \ldots, n. \quad (2.3)$$

**Example 2.3.** In Lagrange interpolation $X = \mathcal{P}_n$ and $L_i f = f(t_{i-1})$, $i = 1, 2, \ldots, n+1$, where $t_0, t_1, \ldots, t_n$ are distinct points. Choosing $x_i = t^{i-1}$ as a basis for $X$

$$\det(L_i x_j) = \begin{vmatrix} 1 & t_0 & t_0^2 & \cdots & t_0^n \\ \vdots & & & & \vdots \\ 1 & t_n & t_n^2 & \cdots & t_n^n \end{vmatrix}.$$

This determinant, known as Vandermonde's determinant, is nonzero if all the $t_i$ are distinct. Hence the Lagrange interpolation problem always has a unique solution.

**Example 2.4.** Let $t_0, t_1, \ldots, t_n$ be distinct points and consider the $N+1$ functionals

$$f(t_0), f'(t_0), \ldots, f^{(m_0)}(t_0)$$

$$f(t_1), f'(t_1), \ldots, f^{(m_1)}(t_1)$$

$$\vdots \qquad\qquad \vdots$$

$$f(t_n), f'(t_n), \ldots, f^{(m_n)}(t_n)$$

where $N = m_0 + m_1 + \cdots + m_n + n$. In $\mathcal{P}_N$ the interpolation with these functionals has a unique solution (Davis, 1963, p. 29). Thus, we can always find a polynomial whose value and derivatives up to a certain order take on prescribed values at distinct points. We will call this a *general Hermite interpolation*.

**Example 2.5.** Let $X$ be the three-dimensional space spanned by 1, cost, sint and let

$$L_1 f = f(-\pi/2), \quad L_2 f = f(0), \quad L_3 f = f(\pi/2).$$

Then the Gram determinant is

$$\begin{vmatrix} 1 & 0 & -1 \\ 1 & 1 & 0 \\ 1 & 0 & 1 \end{vmatrix} = 2.$$

Therefore, we can always find a unique function

$$f(t) = a_0 + a_1 \cos t + a_2 \sin t$$

taking on given values at $-\pi/2$, 0 and $\pi/2$. This is a simple example of *trigonometric interpolation*. More generally, let $X$ be the space spanned by $1, \cos t, \sin t, \cos 2t, \sin 2t, \ldots, \cos nt, \sin nt$. This space has dimension $2n+1$ and for the functionals

$$L_i f = f(t_i), \qquad -\pi \le t_0 < t_1 < \cdots < t_{2n} < \pi$$

the interpolation problem is uniquely solvable (Davis, 1963, p. 30).

**Example 2.6.** The moment problem of finding a function $f(t)$ such that

$$\int_0^1 f(t)\, dt = m_0$$
$$\int_0^1 tf(t)\, dt = m_1$$
$$\vdots$$
$$\int_0^1 t^n f(t)\, dt = m_n$$

does not have a unique solution in $C[0,1]$. If we restrict $f$ to $\mathcal{P}_n$, then the interpolation problem with

$$L_i f = \int_0^1 t^{i-1} f(t)\, dt$$

does have a unique solution. Using $1, t, \ldots, t^n$ as a basis for $\mathcal{P}_n$ the Gram determinant is

$$\begin{vmatrix} 1 & 1/2 & 1/3 & \cdots & 1/(n+1) \\ 1/2 & 1/3 & & \cdots & 1/(n+2) \\ \vdots & & & & \vdots \\ 1/(n+1) & & & \cdots & 1/(2n+1) \end{vmatrix} \neq 0.$$

**Example 2.7.** Not every interpolation problem is necessarily solvable. Suppose we want to find a second degree polynomial whose values at $-1$ and $1$ and whose derivative at $0$ are given. With $L_1 f = f(-1)$, $L_2 f = f(1)$, $L_3 f = f'(0)$, and $x_i = t^{i-1}$ we have

$$\det(L_i x_j) = \begin{vmatrix} 1 & -1 & 1 \\ 1 & 1 & 1 \\ 0 & 1 & 0 \end{vmatrix} = 0$$

and the problem cannot have a unique solution. It may, however, have an infinity of solutions.

**DEFINITION 2.2.** Let $f_1(t), f_2(t), \ldots, f_n(t)$ be functions defined on some subset $S$ of $R^m$. If

$$\det(f_i(t_j)) \neq 0 \qquad (2.4)$$

for all distinct points $t_j$, $j = 1, 2, \ldots, n$ in $S$, then the set of functions is said to be *unisolvent* on $S$.

Unisolvency means that the special interpolation problem

$$\sum_{i=1}^{n} a_i f_i(t_j) = w_j, \qquad j = 1, 2, \ldots, n,$$

can always be solved uniquely. Important examples of unisolvent sets are polynomials, which are unisolvent on any finite segment of the real line, and trigonometric functions (Example 2.5), which are unisolvent on $[-\pi, \pi)$, but not on $[-\pi, \pi]$.

Equation (2.4) is called the *Haar condition*, and a unisolvent set of continuous functions is referred to as a *Chebyshev system*.

**Exercises 2.1**

1. Let $f(t) \in C^{(1)}[a,b]$. Show that for any $\epsilon > 0$ there exists a polynomial $P_n(t)$ such that

$$\|f - P_n\|_\infty \leq \epsilon$$
$$\|f' - P_n'\|_\infty \leq \epsilon.$$

2. Show that if $\dim(X) = \infty$, then $\dim(X^*) = \infty$.
3. For $X = \mathcal{P}_1$ show that the functionals in Example 2.2 are linearly dependent.
4. Show that one can always find a unique second degree polynomial $P_2(t)$ such that

$$P_2(0) = w_1, \quad P_2(1) = w_2, \quad \int_0^1 P_2(t)\,dt = w_3$$

for arbitrary $w_1, w_2, w_3$.
5. Find a third degree polynomial $P_3(t)$ such that

$$P_3(0) = 1, \quad P_3(1) = -1, \quad P_3'(0) = 1, \quad P_3''(0) = 0.$$

6. Show that the set $1, \cos t, \sin t$ is not unisolvent on $[-\pi, \pi]$.
7. Let $X$ be the space spanned by $1, e^t, e^{-t}, 0 \leq t \leq 1$. Show that the linear functionals defined in Example 2.2 are linearly independent in $X^*$.
8. Let $f_1, f_2, \ldots, f_n$ be a Chebyshev system on $S$. Show that there cannot exist a function of the form

$$f(t) = \sum_{i=1}^{n} a_i f_i(t),$$

not identically zero, which has more than $n-1$ zeros in $S$. State and prove a converse for this result.

9. Prove Theorem 2.2.

10. For $X = \mathcal{P}_3$ and functionals defined by

$$L_1 f = f(t_1), \quad L_2 f = f(t_2)$$
$$L_3 f = f''(t_1), \quad L_4 f = f''(t_2)$$

determine whether the interpolation problem has a unique solution.

## 2.2 POLYNOMIAL INTERPOLATION

The results of the previous section have given us a simple condition for the existence and uniqueness of the solution of the general interpolation problem and, through (2.3), a method for computing that solution. However, more is needed if interpolation is to be useful in the construction of numerical algorithms. A more explicit form of the solution is desirable to facilitate formal manipulations, and if interpolation is to be used to approximate functions one needs to consider the question of convergence and determination of error bounds. It is not possible to answer these questions for the general case, but fortunately, a great deal of theory exists for the most important case of polynomial interpolation. We briefly summarize the major results from this theory.

To approximate a function by polynomial interpolation we proceed as follows: Given a function $f(t)$ and functionals $L_1, L_2, \ldots, L_{n+1}$ we try to find a polynomial $P_n(t)$ such that

$$L_i(P_n) = L_i f, \quad i = 1, 2, \ldots, n+1.$$

This is of course a specific case of the general interpolation problem, so that in principle it can be solved using (2.3) with $x_i = t^{i-1}$ and $w_i = L_i f$. In certain simple cases, however, it is possible to obtain an explicit formula for the solution. For example, the Lagrange interpolation polynomial has the very simple representation

$$P_n(t) = \sum_{j=0}^{n} l_j(t) f(t_j) \tag{2.5}$$

where the $l_j(t)$ are the so-called *fundamental polynomials*

$$l_j(t) = \frac{(t-t_0)(t-t_1)\ldots(t-t_{j-1})(t-t_{j+1})\ldots(t-t_n)}{(t_j-t_0)(t_j-t_1)\ldots(t_j-t_{j-1})(t_j-t_{j+1})\ldots(t_j-t_n)}, \quad j = 0, 1, \ldots, n. \tag{2.6}$$

Clearly, $l_j(t_i) = \delta_{ij}$, so that $P_n(t_i) = f(t_i)$, as required. Also, the polynomials $l_j(t)$ are linearly independent and therefore form a basis for $\mathcal{P}_n$.

Equation (2.5) is the *Lagrange form* of the interpolating polynomial. Various other (equivalent) forms, such as the Newton's divided-difference representation, can be found in most texts on numerical analysis. The Lagrange form, while not always the most convenient for the explicit computation of $P_n(t)$, is very suitable for formal manipulation.

Another well-known case is the *simple Hermite interpolation*, which is a special instance of Example 2.4 with $m_0 = m_1 = \cdots = m_n = 1$ (sometimes called osculatory interpolation). With $N = 2n + 1$

$$P_{2n+1}(t) = \sum_{j=0}^{n} \left[ 1 - 2l_j'(t_j)(t - t_j) \right] l_j^2(t) f(t_j)$$

$$+ \sum_{j=0}^{n} (t - t_j) l_j^2(t) f'(t_j). \tag{2.7}$$

It is easily verified that $P_{2n+1}(t_i) = f(t_i)$ and $P_{2n+1}'(t_i) = f'(t_i)$. However, the general Hermite interpolation problem of Example 2.4 does not have such a simple explicit representation.

The next theorem, which is of fundamental importance in the analysis of numerical algorithms, gives us a way of bounding the error in polynomial interpolation.

**THEOREM 2.5.** Let $f(t) \in C^{(N+1)}[a,b]$ and let $P_N(t)$ be the full Hermite interpolating polynomial of Example 2.4, that is,

$$P_N(t_0) = f(t_0), P_N'(t_0) = f'(t_0), \ldots, P_N^{(m_0)}(t_0) = f^{(m_0)}(t_0)$$

$$\vdots$$

$$P_N(t_n) = f(t_n), P_N'(t_n) = f'(t_n), \ldots, P_N^{(m_n)}(t_n) = f^{(m_n)}(t_n)$$

with $N = m_0 + m_1 + \cdots + m_n + n$. We can assume without loss of generality that $a \le t_0 < t_1 < \cdots < t_n \le b$. Then, for $t \in [t_0, t_n]$,

$$f(t) - P_N(t) = \frac{(t - t_0)^{m_0+1}(t - t_1)^{m_1+1} \cdots (t - t_n)^{m_n+1}}{(N+1)!} f^{(N+1)}(\xi), \tag{2.8}$$

where $t_0 \le \xi \le t_n$, and $\xi$ depends on $t$.

PROOF. From Rolle's theorem it follows that if a continuous and differentiable function $f(t)$ has $n$ zeros in $[a,b]$, then $f'(t)$ has at least $n-1$

zeros in $[a,b]$. Furthermore, if $k$ of these zeros have multiplicities greater than one, then $f'(t)$ has at least $n-1+k$ zeros in $[a,b]$. Every zero of multiplicity $k>1$ of $f(t)$ is a zero of multiplicity $k-1$ of $f'(t)$. (Proof of these assertions is left as an exercise.)

Consider now the function

$$W(s) = f(s) - P_N(s)$$

$$- (s-t_0)^{m_0+1}\ldots(s-t_n)^{m_n+1}\frac{f(t)-P_N(t)}{(t-t_0)^{m_0+1}\ldots(t-t_n)^{m_n+1}}.$$

Then $W(s)$ has a zero at $s=t$, and zeros of multiplicity $m_i+1$ at $s=t_i$. Let $\nu_i$ denote the number of zeros of $W^{(i)}(s)$ in $[t_0,t_n]$ and define

$$s(i,j) = 1, \quad \text{if } j > i$$
$$= 0, \quad \text{if } j \leq i.$$

Then

$$\nu_0 = n+2,$$
$$\nu_1 = \nu_0 - 1 + s(0,m_0) + s(0,m_1) + \cdots + s(0,m_n),$$
$$\nu_2 = \nu_1 - 1 + s(1,m_0) + s(1,m_1) + \cdots + s(1,m_n)$$
$$= n + s(0,m_0) + s(1,m_0) + \cdots + s(0,m_n) + s(1,m_n),$$

and finally,

$$\nu_{N+1} = n+1-N+[s(0,m_0) + s(1,m_0) + \cdots + s(N,m_0)]$$

$$\vdots \qquad\qquad\qquad\qquad \vdots$$

$$+[s(0,m_n) + s(1,m_n) + \cdots + s(N,m_n)].$$

But $s(0,m_i) + s(1,m_i) + \cdots + s(N,m_i) = m_i$ so that $\nu_{N+1} = 1$ and $W^{(N+1)}(s)$ has at least one zero in $[t_0,t_n]$, say at $s=\xi$. Differentiating $W(s)$ $N+1$ times and putting $s=\xi$ we get

$$W^{(N+1)}(\xi) = f^{(N+1)}(\xi) - (N+1)!\frac{f(t)-P_N(t)}{(t-t_0)^{m_0+1}\ldots(t-t_n)^{m_n+1}} = 0$$

which proves (2.8). ∎

If $f(t) \in C^{(\infty)}[a,b]$ and if the derivatives of $f(t)$ are uniformly bounded, that is, $\sup_n(\max_{a \leq t \leq b}|f^{(n)}(t)|) \leq K < \infty$, then it follows from (2.8) that

$$P_N(t) \to f(t) \qquad \text{as } n \to \infty,$$

but for less well-behaved functions this is not necessarily true. Even $f \in C^{(\infty)}$ does not always guarantee convergence.

**Example 2.8.** For

$$f(t) = \frac{1}{1+t^2} \quad \text{on } [-a,a]$$

it is known that Lagrange interpolation, using equally spaced points $t_i$, converges to $f(t)$ only in the interval $-3.63 \leq t \leq 3.63$. Outside the interval, the approximating polynomial diverges [this is a classical example due to Runge; see (Davis, 1963, p. 78)]. Not only is convergence a problem, but even if we do have convergence, the rate may be quite slow. Below we give the values of

$$\epsilon = \max_{-a \leq t \leq a} \left| P_N(t) - \frac{1}{1+t^2} \right|$$

for $a = 2$.

| $N$ | 7 | 11 | 15 |
|---|---|---|---|
| $\epsilon$ | $6.0 \times 10^{-2}$ | $4.4 \times 10^{-2}$ | $3.8 \times 10^{-2}$. |

While we can get better results for this particular case by choosing different interpolation points, such as the zeros of the Chebyshev polynomials, the example indicates that there are certain difficulties associated with polynomial interpolation. The general situation is rather complicated, but we can get some insight into the difficulties from the following two theorems. For the purpose of these theorems let $\Delta$ be a triangular array of points

$$
\begin{array}{llll}
t_{00} & & & \\
t_{10} & t_{11} & & \\
t_{20} & t_{21} & t_{22} & \\
\vdots & & & \\
t_{n0} & t_{n1} & \cdots & t_{nn} \\
\vdots & & &
\end{array}
$$

and let $\{P_n(t)\}$ be a sequence of polynomials generated by Lagrange interpolation at the points $a \leq t_{n0} < t_{n1} < \cdots < t_{nn} \leq b$.

**THEOREM 2.6.** For any given triangular set $\Delta$ there exists a continuous function $f(t)$ such that the sequence $\{P_n(t)\}$ does not converge uniformly of $f(t)$.

**THEOREM 2.7.** Given any $f \in C[a,b]$ there exists a $\Delta$ such that $\{P_n(t)\} \to f(t)$ uniformly.

The proofs of these two theorems may be found in Rivlin (1969). Theorem 2.6 indicates that there are no universally effective interpolation schemes, involving function values only. While Theorem 2.7 represents a more positive result it is generally not known what the appropriate $\Delta$ is. The question of convergence is closely connected with the location of the singularities of the extension of $f(t)$ into the complex plane. Some elegant theorems on this subject are given in Davis (1963), where the reader will also find discussions of the convergence of more general schemes, such as Hermite-type interpolation.

We do not pursue this subject, but our discussion does demonstrate certain difficulties associated with function approximation by polynomial interpolation. Furthermore, even when the process is justifiable theoretically, the practical construction of $P_n(t)$ for large $n$ is a tedious process. For these reasons interpolating polynomials of high degree are of limited usefulness in numerical analysis. It is usually much more convenient to split the given interval into smaller subintervals and use a polynomial of relatively low degree over each subinterval. From (2.8) we see that, keeping $N$ fixed, we can get arbitrarily good accuracy by making the subintervals sufficiently small.

### Exercises 2.2

1. Verify that $P_{2n+1}$ in (2.7) satisfies the required interpolation conditions.
2. Prove the assertions made in the first part of the proof of Theorem 2.5.
3. Let $f \in C^{(3)}[0,1]$. Find $P_2(t)$ such that

$$P_2(0) = f(0), \quad P_2(1/2) = f(1/2), \quad P_2(1) = f(1).$$

Obtain a bound on $|f(t) - P_2(t)|$.

4. Let $f \in C^{(4)}[-a,a]$ and let $P_3(t)$ be the polynomial satisfying

$$P_3(-a) = f(-a), \quad P_3(a) = f(a), \quad P_3(0) = f(0), \quad P_3'(0) = f'(0).$$

Show that

$$|f(t) - P_3(t)| \leq \frac{a^4}{96} M_4,$$

where

$$M_4 = \max_{-a \le t \le a} |f^{(4)}(t)|.$$

5. Find an explicit representation for $P_3(t)$ in the preceding exercise of the form

$$P_3(t) = p_1(t)f(a) + p_2(t)f(-a) + p_3(t)f(0) + p_4(t)f'(0).$$

Hint: $p_1(t)$ is a polynomial of degree 3 satisfying $p_1(a)=1$, $p_1(-a)=0$, $p_1(0)=0$, $p_1'(0)=0$, and so on.

6. Use the hint in the previous exercise to find a representation for the Hermite interpolation polynomial with $m_i$ either zero or one.

7. Show that the fundamental polynomials $l_j(t)$ are linearly independent.

8. Let $f \in C^{(n+1)}[0,1]$ with $M_{n+1} = \max|f^{(n+1)}(t)|$ and let $P_n(t)$ be the Lagrange interpolation polynomial on the equidistant points $t_i = ih$, $i = 0, 1, \ldots, n$, $nh = 1$. Show that

$$|f(t) - P_n(t)| \le h^{n+1}M_{n+1}$$

and

$$|f'(t) - P_n'(t)| \le h^n M_{n+1}.$$

## 2.3  PIECEWISE POLYNOMIAL APPROXIMATIONS AND SPLINES

Let $[a,b]$ be a finite interval and introduce a partition $\pi$ of $[a,b]$ by the points $a \le t_0 < t_1 < \cdots < t_n \le b$. The $t_i$ will be called the *nodes* of the partition. We will assume that the partition is uniform, that is

$$t_{i+1} - t_i = h, \quad t_0 = a, \quad t_n = b,$$

but this is merely a matter of convenience.

A function is said to be a *piecewise polynomial of degree k* on $\pi$ if in each subinterval $[t_i, t_{i+1}]$ it is a $k$th degree polynomial. Thus a piecewise polynomial of degree one is a function consisting of piecewise straight line segments. We now investigate the approximation of functions by piecewise polynomials. The construction and analysis of such approximations follow easily from the results of the previous section.

**Example 2.9.** To construct a piecewise 2nd degree polynomial, introduce the additional points $t_{i+1/2}$ midway between the nodes $t_i$ and $t_{i+1}$. We can then find the second degree polynomial by interpolation on $t_i, t_{i+1/2}, t_{i+1}$. If we denote the resulting piecewise 2nd degree polynomial by $P_{n,2}(t)$ then we have from (2.5) that for $t_i \leq t \leq t_{i+1}$,

$$P_{n,2}(t) = \frac{2}{h^2}\left[(t-t_{i+1})(t-t_{i+1/2})f(t_i) - 2(t-t_i)(t-t_{i+1})f(t_{i+1/2})\right.$$
$$\left. + (t-t_i)(t-t_{i+1/2})f(t_{i+1})\right].$$

If $f(t) \in C^{(3)}[a,b]$, then from (2.8),

$$|f(t) - P_{n,2}(t)| \leq \frac{|(t-t_i)(t-t_{i+1/2})(t-t_{i+1})|}{6} M_3,$$

where $M_3 = \max_{a \leq t \leq b}|f^{(3)}(t)|$. Since $t_i \leq t \leq t_{i+1}$, we have immediately

$$|f(t) - P_{n,2}(t)| \leq \frac{h^3}{6} M_3 = \frac{(b-a)^3}{6n^3} M_3, \qquad (2.9)$$

but this is an overestimate and is easily improved (Exercise 1). We do have

$$\lim_{n \to \infty} \|f - P_{n,2}\|_\infty = 0,$$

with an order of convergence of three.

The construction of piecewise $k$th degree interpolating polynomials is accomplished in an analogous manner, and it follows immediately that, if $f \in C^{(k+1)}[a,b]$, the piecewise $k$th degree interpolating polynomial converges to $f(t)$ with order $k+1$. Thus we can get highly accurate approximations while keeping the degree of the polynomial fairly low. There is one drawback, however. While the approximations are continuous at the nodes, they have generally discontinuous derivatives. In some applications this is of no consequence, but often it is desirable to have smoother approximations. One way to accomplish this is to use Hermite interpolation, matching derivatives as well as function values at the nodes.

**Example 2.10.** Using Hermite interpolation we can construct the piecewise 3rd degree polynomial in each subinterval. For $t_i \leq t \leq t_{i+1}$ we get from (2.7)

$$P_{n,3}(t) = \left(1 + \frac{2(t-t_i)}{h}\right)\left(\frac{t-t_{i+1}}{h}\right)^2 f(t_i)$$
$$+ \left(1 - \frac{2(t-t_{i+1})}{h}\right)\left(\frac{t-t_i}{h}\right)^2 f(t_{i+1})$$
$$+ (t-t_i)\left(\frac{t-t_{i+1}}{h}\right)^2 f'(t_i) + (t-t_{i+1})\left(\frac{t-t_i}{h}\right)^2 f'(t_{i+1}).$$

Now $P_{n,3}(t) \in C^{(1)}$ and for $f(t) \in C^{(4)}$ convergence is of order four.

Through Hermite interpolation we can increase the smoothness of the approximating function, but the final expressions include the derivatives of the function at the nodes, which is often undesirable. For function approximation we may have tabulated values of $f$, but not of $f'$; in more complicated cases where $f$ is an unknown to be determined the introduction of the derivative may increase the complexity of the solution process. It is thus natural to ask whether it is possible to construct piecewise polynomial approximations by merely requiring that the approximating function have certain continuity properties at the nodes without explicitly assigning the values of the derivatives there. We are thus led to consider piecewise polynomial approximations $P_{n,k}$ of degree $k$ such that $P_{n,k} \in C^{(p)}$, with $p \geq 0$, that is, $P_{n,k}$ is continuous and has continuous derivatives up to order $p$. In general we are interested only in the case $p < k$, since for $p = k$ the polynomial is the same in all subintervals. Of particular interest is the case which gives the smoothest approximation, $p = k - 1$.

**DEFINITION 2.3.** A piecewise $k$th degree polynomial which has continuous derivatives up to order $k - 1$ is called a *spline* of degree $k$.*

We are particularly interested in interpolating splines, that is, splines which take on prescribed values at the nodes. There is of course nothing in what we have said so far to guarantee that such an interpolating spline exists, in fact, it is known that certain types of spline interpolation problems have no solution. Some specific values of $k$ have been studied extensively, especially $k = 3$, the *cubic spline*. Because of their practical importance, and in order to exhibit the main features of spline approximation without undue complication, we restrict our discussion to cubic splines.

**DEFINITION 2.4.** Let $a = t_0 < t_1 < \cdots < t_n = b$ be a set of nodes in $[a, b]$. Let $f_i$ be a set of prescribed ordinates at $t_i$. A function $\mathcal{C}(t)$ is called the cubic interpolating spline if

(a) $\mathcal{C}(t_i) = f_i$, $i = 0, 1, \ldots, n$
(b) $\mathcal{C}(t) \in C^{(2)}[a, b]$
(c) In each interval $[t_i, t_{i+1}]$ $\mathcal{C}(t)$ is a polynomial of degree $\leq 3$.

We will use $C_i(t)$, $i = 0, 1, \ldots, n - 1$ to denote $\mathcal{C}(t)$ in the interval $[t_i, t_{i+1}]$.

---

*Some authors, for example, Prenter (1975), use the term spline to denote any piecewise polynomial, but the terminology used here appears to be more standard.

First we need to consider the question of existence and uniqueness for cubic splines. To appreciate some of the problems let us look at a simple example.

**Example 2.11.** Let $n=2$. Then we want to find two third-degree polynomials

$$C_0(t) = a_0 + b_0 t + c_0 t^2 + d_0 t^3$$
$$C_1(t) = a_1 + b_1 t + c_1 t^2 + d_1 t^3$$

such that

$$C_0(a) = f_0,$$
$$C_0(t_1) = f_1 = C_1(t_1),$$
$$C_1(b) = f_2,$$
$$C_0'(t_1) = C_1'(t_1),$$
$$C_0''(t_1) = C_1''(t_1).$$

We have six conditions to be satisfied, but eight undetermined parameters, so the solution, if it exists, cannot be unique. To remove the ambiguity two additional conditions must be imposed.

It turns out that a similar situation exists for cubic splines, in general; to make the solution unique we must impose two extra conditions. These could be chosen more or less arbitrarily, but it is reasonable to ask that certain requirements be satisfied at the ends of the interval. For instance, assigning the derivatives at $a$ and $b$ makes the answer unique.

**THEOREM 2.8.** There exists a unique cubic spline $\mathcal{C}(t)$ such that

$$\mathcal{C}(t_i) = f_i, \qquad i = 0, 1, \ldots, n,$$
$$\mathcal{C}'(t_0) = s_0,$$
$$\mathcal{C}'(t_n) = s_n.$$

PROOF.   Consider $C_i(t)$ defined for $t_i \leq t \leq t_{i+1}$ by

$$
\begin{aligned}
C_i(t) = {} & \left\{ \frac{(t - t_{i+1})^2}{h^2} + \frac{2(t - t_i)(t - t_{i+1})^2}{h^3} \right\} f_i \\
& + \left\{ \frac{(t - t_i)^2}{h^2} - \frac{2(t - t_{i+1})(t - t_i)^2}{h^3} \right\} f_{i+1} \\
& + \frac{(t - t_i)(t - t_{i+1})^2}{h^2} s_i + \frac{(t - t_i)^2(t - t_{i+1})}{h^2} s_{i+1}.
\end{aligned}
\tag{2.10}
$$

We immediately see that

$$C_i(t_{i+1}) = C_{i+1}(t_{i+1}) = f_{i+1}$$
$$C_i'(t_{i+1}) = C_{i+1}'(t_{i+1}) = s_{i+1}.$$

Thus the $C_i$ define a piecewise cubic that is continuous, has continuous derivatives, and takes on preassigned values at the nodes. Since $s_0$ and $s_n$ are given, we need only find $s_i$, $i = 1, 2, \ldots, n-1$ such that the $C_i$ also have continuous second derivatives at the nodes. Differentiating (2.10) twice we have

$$C_i''(t_{i+1}) = \frac{6}{h^2} f_i - \frac{6}{h^2} f_{i+1} + \frac{2}{h} s_i + \frac{4}{h} s_{i+1}$$

$$C_{i+1}''(t_{i+1}) = -\frac{6}{h^2} f_{i+1} + \frac{6}{h^2} f_{i+2} - \frac{4}{h} s_{i+1} - \frac{2}{h} s_{i+2}.$$

Equating these two expressions we get

$$s_i + 4s_{i+1} + s_{i+2} = \frac{3}{h}(f_{i+2} - f_i), \qquad i = 0, 1, \ldots, n-2. \tag{2.11}$$

This linear system of $n-1$ equations for the unknowns $s_1, s_2, \ldots, s_{n-1}$ has a matrix that is diagonally dominant and, by a standard result of matrix theory, the system has a unique solution. Hence $\mathcal{C}(t) = C_i(t)$ for $t_i \leq t \leq t_{i+1}$ is a cubic spline. That there is only one such cubic spline follows immediately, for if there were two, then their difference would be a cubic spline taking on values $f_i = 0$ at all nodes. Equation (2.11) then implies that all $s_i$ are zero, and hence the difference between the two splines is identically zero.

The proof which was carried out here only for uniform partitions can be generalized for unequally spaced $t_i$ (see Rivlin, 1963, p. 106). ∎

Another obvious way of introducing the two extra conditions is to require that

$$\mathcal{C}''(t_0) = \mathcal{C}''(t_n) = 0.$$

With these conditions we obtain the so-called *natural spline*. It is left as an exercise to show that there exists a unique cubic natural spline. A complete discussion of the theoretical and practical aspects of spline interpolation is given in Ahlberg, Nilson, and Walsh (1967). We quote two of the more interesting and useful results.

**THEOREM 2.9.**   If $f(t) \in C^{(4)}[a,b]$, then

$$\max_{1 \le i \le n} \sup_{t_{i-1} < t < t_i} |f^{(p)}(t) - \mathcal{C}^{(p)}(t)| \le K_p n^{-4+p} M_4, \qquad 0 \le p \le 3 \quad (2.12)$$

where $K_0 = 5/384, K_1 = \sqrt{3}/216 + 1/24, K_2 = 5/12, K_3 = 1, M_4 = \|f^{(4)}\|_\infty$.

We see from this that cubic spline not only converges to $f(t)$, but also that its derivatives up to order three converge to the corresponding derivatives of $f$.

**THEOREM 2.10.**   Let $U$ be the set of all functions $u \in C^{(2)}$ such that

$$u(t_i) = f_i,$$
$$u'(t_0) = s_0,$$
$$u'(t_n) = s_n,$$

and let $\mathcal{C}(t)$ be the cubic spline satisfying the same interpolating conditions. Then

$$\int_{t_0}^{t_n} [\mathcal{C}''(t)]^2 dt \le \int_{t_0}^{t_n} [u''(t)]^2 dt \qquad (2.13)$$

for all $u \in U$.

The last theorem tells us that of all interpolating functions satisfying certain end conditions, splines are the smoothest in the sense that the integral of the square of the second derivative is minimized. The term spline comes from the fact that draftsmen sometimes use mechanical splines (thin, flexible rods) to draw curves. The spline is fixed at points in the $x-y$ plane and bends to produce a smooth curve. The integral in (2.13) is proportional to the potential energy of the system; (2.13) then says that the spline minimizes the potential energy and therefore is the stable configuration.

The cubic spline can be constructed by solving the simple tridiagonal system (2.11) and substituting the values of $s_i$ in (2.10). For theoretical purposes, however, it is sometimes convenient to have an explicit representation for the spline, similar to the representation of the simple interpolation polynomial by the fundamental polynomials in (2.5).

We define a set of cubic splines $\{c_j(t)\}$ by

$$c_j(t_i) = \delta_{ij}, \qquad j = 0, 1, \ldots, n+2, \quad i = 0, 1, \ldots, n$$
$$c_j'(t_0) = c_j'(t_n) = 0, \qquad j = 0, 1, \ldots, n$$
$$c_{n+1}'(t_0) = c_{n+2}'(t_n) = 1$$
$$c_{n+1}'(t_n) = c_{n+2}'(t_0) = 0.$$

By Theorem 2.8 such splines exist. If we now take

$$\mathcal{C}(t) = \sum_{j=0}^{n} f_j c_j(t) + s_0 c_{n+1}(t) + s_n c_{n+2}(t), \qquad (2.14)$$

then $\mathcal{C}(t)$ satisfies all the conditions of Theorem 2.8 and is therefore a representation for that spline. The $c_j(t)$ are called the *fundamental* or *cardinal* splines.

Similarly, one can construct a simple representation for the cubic natural splines (Exercise 7).

Spline interpolation produces approximations which have several desirable features. The approximations are piecewise polynomials of low degree which are easily constructed and the individual parts are smoothly connected. The approximations converge and produce accurate results for a large class of functions. Finally, they provide approximations not only to the function, but also to its lower order derivatives. For these reasons, splines, particularly cubic splines, have been used extensively in recent years (Ahlberg, Nilson, and Walsh, 1967; Schultz, 1972; Prenter, 1975).

## Exercises 2.3

1. Improve the error bound (2.9).
2. For $f \in C^{(4)}$, show that the piecewise polynomial $P_{n,3}$ in Example 2.10 converges to $f$ with order 4. Find a bound for the error.
3. How large must $n$ be if $\sin t$ is to be approximated in $[0, \pi/2]$ to an accuracy of $10^{-5}$ by a piecewise second degree polynomial constructed by Lagrange interpolation at equally spaced nodes.
4. On $[0, 1]$ with nodes at $0, 0.5, 1$ construct the *quadratic* spline interpolation. You will find that you need to specify one additional condition; do this by assigning the derivative at 0.
5. Construct the cubic natural spline for the partition of the previous problem.
6. Show that the cubic natural spline interpolation problem always has a unique solution.

7. Construct a representation for the cubic natural spline similar to (2.14).

8. Show that there exists a unique quadratic interpolating spline $Q(t)$ such that

$$Q(t_i) = f(t_i), \qquad i = 0, \ldots, n$$
$$Q'(t_0) = s_0.$$

9. Show that the set of all cubic splines on a fixed partition $\pi$ is a linear space with dimension $n+3$.

10. Suppose we are given the values $f(t_0), f(t_1), \ldots, f(t_n)$ and we wish to construct a cubic interpolating spline of the type described in Theorem 2.8. Since in this case we do not know $s_0$ and $s_n$ we use instead $\hat{s}_0$ and $\hat{s}_n$, the finite difference approximations to $f'(t_0)$ and $f'(t_n)$, respectively. Discuss the convergence of this type of spline interpolation.

## 2.4   BEST APPROXIMATIONS IN INNER PRODUCT SPACES

In approximation by interpolation we required that the approximating function satisfy certain simple requirements, which were chosen in some intuitively reasonable way; for instance, in Lagrange interpolation we asked that the approximation agree with the function at certain selected points. Now, rather than imposing such predetermined conditions, we could simply ask for the best approximation to a given function. We must first clarify what we mean by "best," and in a normed space it is natural to define this in terms of the norm of the difference between the function and its approximation.

**DEFINITION 2.5.**   Let $x$ be an element of a normed linear space $X$, and let $\Phi$ be the set of approximations in $X$. Then $\varphi \in \Phi$ is a best approximation if

$$\|x - \varphi\| \leq \|x - v\|, \qquad \text{for all } v \in \Phi. \tag{2.15}$$

Generally $\Phi$ will be a finite-dimensional subspace of $X$. If $\{\varphi_1, \varphi_2, \ldots, \varphi_n\}$ is a basis for $\Phi$, then the problem of finding the best approximation reduces to finding coefficients $a_1, a_2, \ldots, a_n$ which minimize

$$\|x - (a_1\varphi_1 + a_2\varphi_2 + \cdots + a_n\varphi_n)\|.$$

The theoretical analysis and the practical computation of such best approximations depend rather critically on the choice of the approximating space $\Phi$ as well as on the choice of the norm. In the setting of inner product spaces the results are particularly simple and elegant.

The first question which needs to be answered is whether a unique best approximation does in fact exist. In an i.p.s. this question can be answered affirmatively without much difficulty.

**THEOREM 2.11.** Let $X$ be an i.p.s. and let $\varphi_1, \varphi_2, \ldots, \varphi_n \in X$ be linearly independent. Then for any $x \in X$ there exists a unique best approximation in the sense of Definition 2.5.

**PROOF.**

(a)  Existence. Consider

$$d(a_1, a_2, \ldots, a_n) = \|x - (a_1\varphi_1 + a_2\varphi_2 + \cdots + a_n\varphi_n)\|$$

as a function of $a_1, a_2, \ldots, a_n$. Then

$$\big| \|x - (a_1\varphi_1 + \cdots + a_n\varphi_n)\| - \|x - (a_1'\varphi_1 + \cdots + a_n'\varphi_n)\| \big|$$
$$\leq \|(a_1 - a_1')\varphi_1 + \cdots + (a_n - a_n')\varphi_n\|$$
$$\leq |a_1 - a_1'| \|\varphi_1\| + \cdots + |a_n - a_n'| \|\varphi_n\|.$$

Thus, $d(a_1, \ldots, a_n)$ is a continuous function of $a_1, \ldots, a_n$. Similarly, the function $h(a_1, \ldots, a_n) = \|a_1\varphi_1 + \cdots + a_n\varphi_n\|$ is a continuous function. The unit sphere $S : a_1^2 + \cdots + a_n^2 = 1$ is closed and bounded so $h$ takes on a minimum $m$ there. Furthermore, $m > 0$ because of the linear independence of the $\varphi_i$. For general $a_i$, with $r^2 = a_1^2 + \cdots + a_n^2$

$$h = \|a_1\varphi_1 + \cdots + a_n\varphi_n\| = r \left\| \frac{a_1}{r}\varphi_1 + \cdots + \frac{a_n}{r}\varphi_n \right\|$$
$$\geq mr.$$

Hence,

$$\|x - (a_1\varphi_1 + \cdots + a_n\varphi_n)\| \geq \|a_1\varphi_1 + \cdots + a_n\varphi_n\| - \|x\|$$
$$\geq mr - \|x\|$$

is as large as desired for sufficiently large $a_i$. Hence, the minimum of $d$ can occur only in a sphere of finite radius, and since this sphere is closed and bounded the minimum is attained. This proves the existence of the best approximation.

(b)  Uniqueness. Assume that there are two best approximations $\psi_1$ and $\psi_2$ such that

$$\|x - \psi_1\| = \|x - \psi_2\| = \epsilon.$$

It cannot be true that

$$\|x - (\psi_1 + \psi_2)/2\| < \epsilon,$$

since this contradicts the assumption that $\psi_1$ and $\psi_2$ are best approximations. Therefore, we must have

$$\|(x - \psi_1) + (x - \psi_2)\| = 2\epsilon = \|x - \psi_1\| + \|x - \psi_2\|.$$

From Theorem 1.4 it then follows that $\psi_1 = \psi_2$. ∎

To compute the best approximation we need to find the coefficients $a_i$ which minimize the error. This minimization problem can be formulated rather simply as a set of linear equations.

**THEOREM 2.12.** Let $X$ be an i.p.s. and let $\Phi$ be a subspace of $X$ spanned by $\varphi_1, \varphi_2, \ldots, \varphi_n$. For a given $x \in X$ the best approximation $\varphi \in \Phi$ to $x$ is given by

$$\varphi = \sum_{i=1}^{n} a_i \varphi_i$$

where the coefficients $a_i$ are the solution of the *normal equations*

$$(\varphi_1, \varphi_1) a_1 + (\varphi_2, \varphi_1) a_2 + \cdots + (\varphi_n, \varphi_1) a_n = (x, \varphi_1)$$
$$\vdots \qquad\qquad\qquad\qquad \vdots \qquad\quad (2.16)$$
$$(\varphi_1, \varphi_n) a_1 + (\varphi_2, \varphi_n) a_2 + \cdots + (\varphi_n, \varphi_n) a_n = (x, \varphi_n)$$

PROOF.   Consider

$$\left\| x - \sum_{i=1}^{n} a_i \varphi_i \right\|^2 = \left( x - \sum_{i=1}^{n} a_i \varphi_i, x - \sum_{i=1}^{n} a_i \varphi_i \right).$$

This is a continuous and differentiable function of the $a_i$, so to find the minimum we set the partial derivatives with respect to the $a_i$ to zero. Hence, we must have

$$\frac{\partial}{\partial a_j}(x,x) - 2\frac{\partial}{\partial a_j}\sum_{i=1}^{n} a_i(x,\varphi_i) + \frac{\partial}{\partial a_j}\sum_{i=1}^{n}\sum_{k=1}^{n} a_i a_k(\varphi_i,\varphi_k) = 0$$

or

$$\sum_{i=1}^{n} a_i(\varphi_i,\varphi_j) = (x,\varphi_j), \qquad j = 1,2,\ldots,n.$$

It remains to be shown that (2.16) has a unique solution and that this solution does give a minimum; we will leave this as an exercise. ∎

Normal equations are superficially attractive for the solution of the best approximation problem, but a certain amount of caution is in order.

**Example 2.12.** In $L_2[0,1]$ we can define an inner product by

$$(x,y) = \int_0^1 x(t)y(t)\,dt.$$

For $\varphi_i(t) = t^{i-1}$

$$(\varphi_i,\varphi_j) = \frac{1}{i+j-1}$$

and the matrix associated with the normal equations is the notoriously ill-conditioned Hilbert matrix. For $n > 5$ or 6 the approach through the normal equations becomes computationally unmanageable.

Ideally, we would like to choose the $\varphi_i$ such that the normal equations are well-conditioned and simple. This is certainly the case if

$$(\varphi_i,\varphi_j) = \delta_{ij}$$

in which case we also get the explicit form of the coefficients

$$a_i = (x,\varphi_i).$$

**DEFINITION 2.6.** A sequence of elements $\varphi_1,\varphi_2,\ldots,\varphi_n$ in an i.p.s. is said to be *orthogonal* if

$$(\varphi_i,\varphi_j) = 0, \qquad \text{for } i \neq j. \tag{2.17}$$

The sequence is said to be *orthonormal* if

$$(\varphi_i,\varphi_j) = \delta_{ij}. \tag{2.18}$$

**Example 2.13**

(a)  In $R^n$, with the inner product defined by (1.13), the unit vectors

$$e_i = \begin{bmatrix} 0 \\ \vdots \\ 1 \\ \vdots \\ 0 \end{bmatrix} \leftarrow i\text{th row}$$

are orthonormal.

(b)  In $L_2[-\pi,\pi]$ with inner product

$$(x,y) = \int_{-\pi}^{\pi} x(t)y(t)\,dt$$

the functions

$$1/\sqrt{2\pi}\,, 1/\sqrt{\pi}\ \cos t, 1/\sqrt{\pi}\ \sin t, 1/\sqrt{\pi}\ \cos 2t, 1/\sqrt{\pi}\ \sin 2t, \ldots$$

are an orthonormal sequence.

(c)  In $C[-1,1]$ with inner product

$$(x,y) = \int_{-1}^{1} \frac{x(t)y(t)}{(1-t^2)^{1/2}}\,dt$$

the functions $1/\sqrt{\pi}\,, \sqrt{2/\pi}\ T_n(t)$,  $n=1,2,\ldots$  are orthonormal. Here the $T_n$ are the Chebyshev polynomials

$$T_n(t) = \cos(n\cos^{-1}t).$$

**THEOREM 2.13.**  If $x_1, x_2, \ldots, x_n$ form an orthogonal set, then they are linearly independent.

PROOF.  Assume that

$$\alpha_1 x_1 + \alpha_2 x_2 + \cdots + \alpha_n x_n = 0.$$

Taking the inner product with $x_k$ we have

$$\alpha_k(x_k, x_k) = 0$$

and since $x_k \neq 0$, this implies that $\alpha_k = 0$. Repeating this argument for the other elements we get that $\alpha_1 = \alpha_2 = \cdots = \alpha_n = 0$ and hence the set is linearly independent. ∎

Conversely, if we have any linearly independent set $x_1, x_2, \ldots, x_n$ we can generate an orthonormal set $x_1^*, x_2^*, \ldots, x_n^*$ by taking a linear combination of the $x_i$. This can be done in a number of different ways; a systematic approach is by the Gram–Schmidt process.

**THEOREM 2.14.** Let $x_1, x_2, \ldots, x_n, \ldots$ be a sequence of elements in an i.p.s. such that any finite subset of the $x_i$ is linearly independent. Then we can find a set of coefficients $a_{ij}$ such that

$$x_i^* = \sum_{j=1}^{i} a_{ij} x_j, \qquad i = 1, 2, \ldots \tag{2.19}$$

form an orthonormal set. This set can be constructed by the *Gram–Schmidt orthonormalization* process in which the $x_i^*$ are computed by

$$
\begin{aligned}
&y_1 = x_1, \quad x_1^* = y_1 / \|y_1\|, \\
&y_2 = x_2 - (x_2, x_1^*) x_1^*, \quad x_2^* = y_2 / \|y_2\|, \\
&\vdots \\
&y_n = x_n - \sum_{j=1}^{n-1} (x_n, x_j^*) x_j^*, \quad x_n^* = y_n / \|y_n\|.
\end{aligned}
\tag{2.20}
$$

PROOF.   Since

$$y_n = x_n + \text{linear combination of } (x_1^*, \ldots, x_{n-1}^*),$$

it is obvious that

$$y_n = x_n + \text{linear combination of } (x_1, \ldots, x_{n-1});$$

hence $x_n^*$ has the required form (2.19). Also, we cannot have $\|y_n\| = 0$, since this would imply that

$$0 = y_n = \text{linear combination of } (x_1, \ldots, x_n)$$

contradicting the assumption of linear independence of the $x_i$. Thus, the process cannot break down at any step. Now assume that $x_1^*, x_2^*, \ldots, x_{n-1}^*$

are orthonormal. Then

$$(y_n, x_i^*) = (x_n, x_i^*) - \sum_{j=1}^{n-1} (x_n, x_j^*)(x_j^*, x_i^*)$$

$$= (x_n, x_i^*) - (x_n, x_i^*) = 0, \qquad \text{for } i = 1, 2, \ldots, n-1.$$

Since $\|x_n^*\| = 1$, it follows that the set $x_1^*, x_2^*, \ldots, x_n^*$ is orthonormal. Since the assumption is true for $n = 2$, we have by induction that the generated sequence is orthonormal. ∎

If $X$ is an $n$-dimensional space, then linearly independent elements $x_1, x_2, \ldots, x_n$ are a basis for $X$. If these are orthonormalized, then the $x_i^*$ are again a basis. Representation of elements of $X$ and the determination of the best approximation are very simple with an orthonormal basis. The resulting form is called the *Fourier expansion*.

**THEOREM 2.15.**  Let $X$ be an i.p.s. of dimension $n$, and let $x_1^*, \ldots, x_n^* \in X$ be an orthonormal basis. Then any $x \in X$ can be expressed as

$$x = \sum_{i=1}^{n} \alpha_i x_i^*$$

where

$$\alpha_i = (x, x_i^*). \tag{2.21}$$

PROOF.  This follows directly by taking the inner product of the expansion with $x_i^*$ and using the orthonormality. ∎

**DEFINITION 2.7.**  Let $x_1^*, x_2^*, \cdots$ be an infinite sequence of orthonormal elements in an i.p.s. Then the series

$$\sum_{j=1}^{\infty} (x, x_j^*) x_j^*$$

is called the *Fourier series* for $x$. The finite sum

$$\sum_{j=1}^{n} (x, x_j^*) x_j^*$$

is the *truncated Fourier series*.

Since $(x,x_j^*)x_j^*$ is the orthogonal projection of $x$ onto $x_j^*$, the truncated Fourier series is the orthogonal projection of $x$ onto the subspace spanned by $x_1^*,\ldots,x_n^*$. We now show that the truncated Fourier series is, in fact, the solution of the best approximation problem.

**THEOREM 2.16.** If $x_1^*,x_2^*,\ldots,x_n^*$ are orthonormal, then for any $x \in X$

$$\left\| x - \sum_{i=1}^{n} (x,x_i^*)x_i^* \right\| \le \left\| x - \sum_{i=1}^{n} a_i x_i^* \right\|$$

for all possible choices of $a_1,a_2,\ldots,a_n$.

PROOF.

$$\left\| x - \sum_{i=1}^{n} a_i x_i^* \right\|^2 = \left( x - \sum_{i=1}^{n} a_i x_i^*, x - \sum_{i=1}^{n} a_i x_i^* \right)$$

$$= (x,x) - 2 \sum_{i=1}^{n} a_i (x,x_i^*) + \sum_{i=1}^{n} a_i^2$$

$$= (x,x) - \sum_{i=1}^{n} (x,x_i^*)^2 + \sum_{i=1}^{n} \left[ a_i - (x,x_i^*) \right]^2.$$

Since the first two terms are independent of $a_i$ the expression can be minimized by taking $a_i = (x,x_i^*)$, which proves the theorem. ■

From the preceding proof it follows that

$$(x,x) - \sum_{i=1}^{n} (x,x_i^*)^2 \ge 0$$

or

$$\sum_{i=1}^{n} (x,x_i^*)^2 \le \|x\|^2, \tag{2.22}$$

the so-called *Bessel inequality*.
While Bessel's inequality implies that

$$\lim_{n \to \infty} (x,x_n^*) = 0 \tag{2.23}$$

for all $x$, it does not imply that the sequence of truncated Fourier expansions converges to $x$.

**Example 2.14.** The functions $1/\sqrt{\pi}\,\sin kt$, $k = 1, 2, \cdots$ on the interval $[-\pi, \pi]$ with the inner product defined as in Example 2.13(b), form an orthonormal sequence. But since these functions are all odd, the truncated Fourier series obviously cannot converge if $x$ is an even function.

The relation between the question of the convergence of the Fourier series and some of its properties is addressed by the following well-known theorem.

**THEOREM 2.17.** Let $x_1^*, x_2^*, \cdots$ be a sequence of orthonormal elements forming a basis for an i.p.s. $X$. Then

(a) For any $x \in X$ the truncated Fourier series converges to $x$, that is,

$$\lim_{n \to \infty} \left\| x - \sum_{i=1}^{n} (x, x_i^*) x_i^* \right\| = 0. \tag{2.24}$$

(b) The *Parseval identity*

$$\|x\|^2 = \sum_{i=1}^{\infty} (x, x_i^*)^2 \tag{2.25}$$

and the *extended Parseval identity*

$$(x, y) = \sum_{i=1}^{\infty} (x, x_i^*)(y, x_i^*) \tag{2.26}$$

hold for any $x, y \in X$.

(c) If $(x, x_i^*) = 0$ for all $i$, then $x = 0$.

(d) An element is uniquely determined by its Fourier coefficient, that is, if

$$(x, x_i^*) = (y, x_i^*), \qquad \text{for all } i,$$

then $x = y$.

The proof, as well as a more general statement of the theorem, can be found in (Davis, 1963, p. 192).

Part (a) of this theorem states that if the $x_i^*$ are such that any $x \in X$ can be approximated arbitrarily closely by a linear combination of the $x_i^*$, then the Fourier series for $x$ converges to $x$. This is a fairly trivial observation, a more difficult problem is to determine whether the $x_i^*$ are indeed a basis for $X$. Fortunately the cases of primary computational interest are well understood. The Weierstrass theorem implies that orthogonal polynomials on a finite interval are a basis for the space of continuous functions; similarly, trigonometric expansions are covered by the results of classical Fourier theory.

We see from this discussion that the theory of best approximations in an i.p.s. is quite simple and elegant. However, in the more general setting of the normed spaces complications arise. Going back to the proof of Theorem 2.11 we see that no use of inner products was made in proving the existence of a best approximation. Thus a best approximation exists in any normed space. The proof of uniqueness, however, required the use of Theorem 1.4. Any norm for which Theorem 1.4 holds is said to be a *strict* norm, and it follows then that in a normed space with a strict norm the best approximation is unique.

It can be shown by appropriate manipulation of the Minkowski inequality [see (Davis, 1963, p. 132)], that for $1 < p < \infty$ the $L_p$ norm defined by (1.5) is strict. Unfortunately, for the two most interesting cases, $p = 1$ and $p = \infty$, this is not the case. The case $p = 1$ has so far not been used extensively in numerical analysis and we do not discuss it here. Rivlin (1969) gives a brief account of the known results. Best approximations in the maximum norm $p = \infty$, however, are extremely important, and we discuss some of the best known results in the next section.

### Exercises 2.4

1. Show that for polynomials, with $p_{-1} = 0$ and $p_0 = 1$ and an inner product as defined in Example 1.8, the Gram–Schmidt process collapses to a three-term recursion relation

$$p_{n+1}(t) = tP_n(t) - (tP_n, P_n)P_n(t) - (p_n, p_n)^{1/2}P_{n-1}(t)$$

$$P_{n+1}(t) = \frac{p_{n+1}(t)}{(p_{n+1}, p_{n+1})^{1/2}}, \qquad n = 1, 2, \dots.$$

2. Find the least-squares approximation to $e^{-t}$ by a linear combination of $1, t, t^3$ on the interval $[0, 1]$. By a least-squares approximation we mean the best approximation with the norm defined by the inner product in Example 1.8.
   (a)  use the normal equations;
   (b)  use Fourier series and the Gram–Schmidt orthonormalization.

3. Orthonormalize $1, t, t^2$ with respect to the inner product

$$(x,y) = \int_0^1 e^{-t}x(t)y(t)\,dt.$$

4. Complete the proof of Theorem 2.12 by showing that
   (a) The normal equations have a unique solution.
   (b) The solution so obtained minimizes $\|x - \Sigma a_i \varphi_i\|$.
5. Prove the Parseval identities (2.25) and (2.26).
6. Show by example that for $p = 1$ and $p = \infty$ the norm defined by (1.5) is not strict.

## 2.5 BEST APPROXIMATIONS IN THE MAXIMUM NORM

Polynomial approximations are used widely in the construction of computer programs for the evaluation of trigonometric, exponential, and other standard functions. In this context it is natural to ask that the error be minimized pointwise, rather than in some average sense; hence, one looks for best approximations in the maximum norm. Only the existence of such approximations, usually called *Chebyshev* or *min-max* approximations, is guaranteed by the result of the previous section, but it is known that such approximations are unique. Also, a simple characterization of the min-max solution is known, which allows its practical construction. Finally, error bounds have been studied in detail. Many of the proofs required to establish these results are technical and tedious, and are of little interest here. We therefore quote the major theorems with appropriate references to works where the proofs may be found.

**THEOREM 2.18.** Let $f(t) \in C[a,b]$. For fixed $n$ there exists a unique $P_n^* \in \mathcal{P}_n$ such that

$$\max_{a \le t \le b} |P_n^*(t) - f(t)| \le \max_{a \le t \le b} |P_n(t) - f(t)|$$

for all $P_n \in \mathcal{P}_n$.

PROOF. This is a special case of a theorem given in (Davis, 1963, p. 143). ∎

While this theorem guarantees the existence and uniqueness of the min-max approximation, more information is needed for the actual construction of such an approximation. This information is contained in a famous theorem due to Chebyshev.

**THEOREM 2.19 CHEBYSHEV EQUI-OSCILLATION THEOREM.**
For $f(t) \in C[a,b]$ let $P_n^*(t)$ be the min-max approximation polynomial of degree $n$. Set

$$\epsilon(t) = f(t) - P_n^*(t)$$
$$E_m = \max_{a \le t \le b} |\epsilon(t)|.$$

Then there exist at least $n+2$ distinct points $a \le t_1 < t_2 < \cdots < t_{n+2} \le b$ such that

$$|\epsilon(t_i)| = E_m, \qquad i = 1, 2, \ldots, n+2 \tag{2.27}$$

and

$$\epsilon(t_i) = -\epsilon(t_{i+1}), \qquad i = 1, 2, \ldots, n+1. \tag{2.28}$$

Conversely, if there are $n+2$ distinct points for which (2.27) and (2.28) are satisfied and

$$|\epsilon(t)| \le E_m, \qquad \text{for all } t \in [a,b],$$

then $P_n^*(t)$ is the min-max approximation.

PROOF.   See (Davis, 1963, pp. 149–152).                                    ■

**Example 2.15.**   Find the min-max approximation polynomial of degree one to $\sqrt{t}$ in $[a,b]$. We write

$$P_1^*(t) = c_1 + c_2 t$$

then

$$\epsilon(t) = \sqrt{t} - c_1 - c_2 t.$$

Since

$$\frac{d\epsilon}{dt} = \frac{1}{2\sqrt{t}} - c_2,$$

there is only one interior extremum at $t = 1/(4c_2^2)$. As there have to be at least three extrema of $\epsilon(t)$, the other two must be at the ends of the interval. Thus, we have

$$\sqrt{a} - c_1 - c_2 a = E_m$$

$$\frac{1}{2c_2} - c_1 - \frac{1}{4c_2} = -E_m$$

$$\sqrt{b} - c_1 - c_2 b = E_m.$$

This nonlinear system is easily solved and we obtain

$$c_1 = \frac{1}{2}\left[\sqrt{a} - \frac{a}{\sqrt{a} + \sqrt{b}} + \frac{\sqrt{a} + \sqrt{b}}{4}\right],$$

$$c_2 = \frac{1}{\sqrt{a} + \sqrt{b}},$$

$$E_m = \sqrt{b} - c_1 - c_2 b.$$

For higher $n$ it is usually not possible to solve the resulting nonlinear system directly, but various algorithms are known by which min-max approximations can be constructed numerically. Basically, one starts with some polynomial and then iteratively adjusts the coefficients to satisfy the equi-oscillation conditions more and more closely. For more details see (Rivlin, 1969, pp. 40–43).

**THEOREM 2.20.** Let $f(t) \in C[a,b]$ and let $P_n^*(t)$, $n = 0, 1, \cdots$ be the sequence of min-max approximations to $f(t)$. Then

$$\lim_{n \to \infty} P_n^*(t) = f(t).$$

PROOF. This is an immediate consequence of the Weierstrass theorem. For any $\epsilon > 0$ there exists for sufficiently large $n$ a polynomial $P_n(t)$ such that

$$\max_{a \le t \le b} |f(t) - P_n(t)| \le \epsilon.$$

Since $P_n^*(t)$ is the best approximation in the maximum norm the same inequality must also be satisfied for it.  ∎

Error bounds for min-max approximations were given in a set of theorems by Jackson, of which the following is a special case.

**THEOREM 2.21.** Let $f(t) \in C^{(k+1)}[-1,1]$ and let

$$\max_{-1 \le t \le 1} |f^{(k+1)}(t)| = M_{k+1}. \tag{2.29}$$

If $P_n^*(t)$ is the min-max approximation to $f(t)$, then for $n > k$

$$|f(t) - P_n^*(t)| \le \frac{6^{k+1}e^k}{n^k(n-k)(1+k)} M_{k+1}. \tag{2.30}$$

PROOF.   See (Rivlin, 1969, p. 23).                                                    ∎

If $f(t)$ is sufficiently smooth ($k$ large) convergence becomes very rapid for large $n$; unfortunately, for the range of $n$ which is of most practical significance ($n < 20$), (2.30) yields totally unrealistic and useless error bounds. If $f(t)$ is sufficiently differentiable, a more useful result can be obtained without much difficulty.

**THEOREM 2.22.** Let $f(t) \in C^{(n+1)}[-1,1]$ and let $M_{n+1}$ be defined by (2.29). Then

$$|f(t) - P_n^*(t)| \le \frac{1}{2^n(n+1)!} M_{n+1}. \tag{2.31}$$

PROOF.   Let $z_1, z_2, \ldots, z_{n+1}$ be the roots of the Chebyshev polynomial $T_{n+1}$. Let $P_n(t)$ be the polynomial which interpolates $f(t)$ at the points $z_i$. Then from (2.8),

$$f(t) - P_n(t) = \frac{r_n(t)}{(n+1)!} f^{(n+1)}(\xi)$$

where $r_n(t) = (t - z_1)(t - z_2) \cdots (t - z_{n+1})$. Clearly $r_n(t)$ is a multiple of $T_{n+1}(t)$, in fact, it is easily shown (Exercise 4) that $r_n(t) = T_{n+1}(t)/2^n$. Since $|T_{n+1}(t)| \le 1$, the bound (2.31) follows.                                        ∎

If the bounds on the higher derivatives do not increase too rapidly, then (2.31) is a realistic error bound. In fact, for simple cases, such as $f(t) = \sin t$, the polynomial obtained by interpolation at the zeros of $T_{n+1}$ is quite close to the actual min-max approximation.

**Exercises 2.5**

1. If $f(t)$ is symmetric about the origin, show that the min-max polynomial approximation to $f(t)$ is symmetric also.

2. Find the error bounds for the min-max approximation of sin t on $[-1, 1]$ by polynomials of degree 10, 20, and 30 using (2.30) and (2.31) and compare the results.

3. Find the min-max linear approximation to $e^t$ in $[0, 1]$.

4. (a) Show that the Chebyshev polynomials $T_n(t) = \cos(n \cos^{-1} t)$ satisfy the recursion relation

$$T_n(t) = 2t T_{n-1}(t) - T_{n-2}(t), \qquad n = 2, 3, \ldots$$

and hence, that $T_n(t)$ is indeed a polynomial of degree $n$.

(b) Show that the Chebyshev polynomials form an orthogonal sequence for the inner product defined by

$$(f, g) = \int_{-1}^{-1} (1 - t^2)^{-1/2} f(t) g(t) \, dt.$$

(c) Show that $T_n(t) = 2^{n-1} t^n + \text{lower powers of } t$, and hence, that in Theorem 2.22 $r_n(t) = T_{n+1}(t)/2^n$.

(d) Let $\tilde{\mathscr{P}}_n$ be the set of all polynomials of degree $n$ of the form $\tilde{p}(t) = t^n + \text{lower powers}$. Show that

$$\frac{1}{2^{n-1}} \max_{-1 \le t \le 1} |T_n(t)| \le \max_{-1 \le t \le 1} |\tilde{p}(t)|, \qquad \text{for all } \tilde{p} \in \tilde{\mathscr{P}}_n.$$

## 2.6 APPROXIMATIONS IN SEVERAL VARIABLES

Approximation of functions of several independent variables is a topic of considerable difficulty. While some of the previous results, such as approximation in an i.p.s., are general and apply to more than one dimension, others, such as those in Section 2.5, are limited to functions of one variable. Apart from any theoretical difficulty, there are also practical problems in constructing approximations in several variables.

Some of the theorems we established for functions of one variable can be generalized. The possibility of approximating continuous functions of $n$ variables by polynomials (in $n$ variables) is guaranteed by the *Stone–Weierstrass* theorem (Davis, 1963, p. 122) which is an extension of the classical Weierstrass theorem. But it must be realized that not every result can be carried over to the multivariate case quite as nicely, and that there exist difficulties for $n \ge 2$ that are not present in the one-dimensional case. For example, we indicated in Section 2.1 that for $n = 1$ polynomials were unisolvent. This is no longer true for $n \ge 2$; in fact, a theorem of *Haar* states that there are no unisolvent sets of continuous functions for $n \ge 2$ (Davis, 1963, p. 32). Thus, existence and uniqueness questions have to be considered carefully.

Still, from a practical standpoint, it is often possible to generalize and use the results from one dimension to construct multivariate approximations. We give several examples to demonstrate how this is done in the approximation of functions of two variables by interpolation.

**Example 2.16.** Let $(t_1,s_1),(t_2,s_2),(t_3,s_3)$ be three distinct points in $R^2$. Find a polynomial of the form

$$P(t,s) = a_0 + a_1 t + a_2 s$$

taking on values $f_1, f_2, f_3$, respectively, at these three points. Using the technique of Section 2.1 we obtain the system

$$a_0 + a_1 t_1 + a_2 s_1 = f_1$$
$$a_0 + a_1 t_2 + a_2 s_2 = f_2$$
$$a_0 + a_1 t_3 + a_2 s_3 = f_3$$

from which the unknowns can be determined.

The construction suggested in the above example can, in principle, be used to find interpolating polynomials of higher degree. However, for such a problem to have a unique solution it is necessary that the resulting matrix be nonsingular; unfortunately, unlike the one-dimensional case, this is not always so. Even in the very simple instance of Example 2.16 there are exceptional cases (see Exercise 1). Thus, a certain amount of caution has to be exercised in multidimensional interpolation. There is one case when everything is quite simple; this happens when the interpolating points form a rectangular grid. Here it is possible to generalize most of the results of Section 2.2 by repeated one-dimensional interpolation.

**THEOREM 2.23.** Let $f(t,s)$ be a given function and let $t_0, t_1, \ldots, t_n$ and $s_0, s_1, \ldots, s_m$ distinct points along the $t$ and $s$ axes, respectively. Then there exists a unique polynomial $P(t,s)$ of degree $n$ in $t$ and of degree $m$ in $s$ such that

$$P(t_i,s_j) = f(t_i,s_j) = f_{ij}, \qquad i = 0,1,\ldots,n, \quad j = 0,1,\ldots,m. \quad (2.32)$$

**PROOF.** The existence of such a polynomial can be shown by construction. Consider

$$P(t,s) = \sum_{i=0}^{n} \sum_{j=0}^{m} l_i(t) l_j(s) f_{ij}$$

where the $l_i$ are the fundamental polynomials defined by (2.6). Then $l_i(t_k)l_j(s_v) = \delta_{ik}\delta_{jv}$ and the interpolating conditions are satisfied, proving the existence of an interpolating polymial.

To prove uniqueness, let us write the polynomial in the form

$$P(t,s) = \sum_{i=0}^{n} \sum_{j=0}^{m} a_{ij} t^i s^j$$

and assume that there exists another interpolating polynomial

$$\hat{P}(t,s) = \sum_{i=0}^{n} \sum_{j=0}^{m} b_{ij} t^i s^j.$$

Then since both satisfy the interpolating conditions, we have

$$\sum_{i=0}^{n} \sum_{j=0}^{m} (a_{ij} - b_{ij}) t_k^i s_v^j = 0, \qquad \begin{matrix} k=0,1,\dots,n \\ v=0,1,\dots,m. \end{matrix}$$

If we now set

$$c_{iv} = \sum_{j=0}^{m} (a_{ij} - b_{ij}) s_v^j$$

then since the matrix $(t_k^i)$ is a Vandermonde matrix, we must have $c_{iv} = 0$ for all $i$ and $v$. Repeating this argument with $(s_v^j)$ we find that

$$a_{ij} = b_{ij}, \qquad \text{for all } i \text{ and } j. \quad \blacksquare$$

In a similar fashion we can extend the error bounds of Section 2.2. In the simplest form we get

**THEOREM 2.24.** Let $f(t,s)$ be in $C^{(n+1)}$ with respect to $t$ and in $C^{(m+1)}$ with respect to $s$. Also, take uniformly spaced grid points in each direction, that is, $t_{i+1} - t_i = h_t$ and $s_{i+1} - s_i = h_s$. Then

$$|f(t,s) - P(t,s)| = O(h_t^{n+1}) + O(h_s^{m+1}). \tag{2.33}$$

PROOF. We shall only sketch the arguments, leaving the details to the reader. We define a function $\varphi(t,s)$ by

$$\varphi(t,s) = \sum_{i=0}^{n} l_i(t) f(t_i,s).$$

Then for fixed $s, \varphi(t,s)$ is a polynomial of degree $n$ in $t$, so that from Theorem 2.5 and Exercise 8 in Section 2.2 we see that

$$|\varphi(t,s) - f(t,s)| = O(h_t^{n+1}).$$

Also,

$$P(t,s) = \sum_{j=0}^{m} l_j(s)\varphi(t,s_j)$$

so that

$$|P(t,s) - \varphi(t,s)| = O(h_s^{m+1}).$$

Putting these results together we obtain (2.33)  ∎

A similar approach can be used to construct two-dimensional splines on a rectangular grid. Let $c_i(t)$ be the one-dimensional cubic cardinal spline, defined in Section 2.3. Then

$$S(t,s) = \sum_{i=0}^{n+2} \sum_{j=0}^{m+2} \beta_{ij} c_i(t) c_j(s) \qquad (2.34)$$

is a two-dimensional spline. With $\beta_{ij} = f_{ij}$, $i = 0, 1, \ldots, n$, $j = 0, 1, \ldots, m$ the interpolation conditions are satisfied. The undetermined parameters can be fixed by prescribing conditions at the boundaries of the region. We can see what types of additional conditions might be required by counting the number of parameters $\beta_{ij}$. There are $(n+3)(m+3)$ of these in (2.34); we impose $(n+1)(m+1)$ conditions by the interpolating conditions, leaving $2m + 2n + 8$ conditions to be assigned. For example, we may require that $S(t,s)$ have the same normal derivatives as $f(t,s)$ at the edge mesh-points and that the second cross-derivatives match at the corners. That is, we require that

$$\frac{\partial S}{\partial t}(t_i, s_j) = \frac{\partial f}{\partial t}(t_i, s_j), \qquad i = 0, n, \quad j = 0, 1, \ldots, m$$

$$\frac{\partial S}{\partial s}(t_i, s_j) = \frac{\partial f}{\partial s}(t_i, s_j), \qquad i = 0, 1, \ldots, n, \quad j = 0, m$$

$$\frac{\partial^2 S}{\partial t \partial s}(t_i, s_j) = \frac{\partial^2 f}{\partial t \partial s}(t_i, s_j), \qquad i = 0, n, \quad j = 0, m.$$

It is known that the spline so generated is unique and that, for equally spaced grid points and sufficiently smooth $f$, the error in the approximation is of order $h^4$.

The above discussion indicates that either piecewise polynomials or splines provide an efficient method for approximating functions of two variables provided the region of interest can be subdivided into rectangles. Difficulties arise when this is not so, since curved boundaries cannot be closely approximated by rectangles unless the mesh-size is very small. For this reason there is much current interest in the construction of piecewise polynomials and splines on triangular regions or regions with curved boundaries. The subject is technically complicated and we must refer the reader to the literature on it.

Approximation of functions of several variables is a very new topic and many of the results are of recent origin. As a result it is difficult to find good discussions of these matters in standard texts. An exception is the very readable introduction provided in (Prenter, 1975, ch. 5). Further information can be found in (Rice, 1969) and (Birkhoff and deBoor, 1965).

### Exercises 2.6

1. It is possible for the system in Example 2.16 not to have a unique solution? Give a geometrical interpretation of the exceptional cases.
2. Find the bilinear function

$$P(t,s) = a_0 + a_1 t + a_2 s + a_3 ts$$

such that $P(0,0)=f_{00}, P(1,0)=f_{10}, P(0,1)=f_{01}, P(1,1)=f_{11}$.
3. Find the second degree polynomial which interpolates a function on the grid $t_i = 0, 1, 2$, $s_i = 0, 1, 2$.
4. Verify that $S(t,s)$ in (2.34) satisfies $S(t_i, s_j) = f_{ij}$. Show that $S(t,s)$ is twice continuously differentiable.

# II

# THEORETICAL ASPECTS OF COMPUTATIONAL MATHEMATICS

Having provided the necessary framework we are now ready to undertake the development of a general theory of approximate computation. In Chapter 3 we investigate the most important direct problem, namely, numerical integration. The treatment here consists of a quite elementary application of the ideas contained in the first two chapters, and while it perhaps sheds no new light on the major problems of numerical integration (which are primarily practical ones) it does show the structure underlying most quadrature algorithms. In Chapter 4 we present the topic central to numerical analysis, the approximate solution of the linear inverse problem. We discuss the theoretical aspects of the construction of various algorithms and so are led to the important theorems relating the concepts of consistency, stability, and convergence. A further development deals with the justification of certain extrapolation procedures frequently used in practice. A brief discussion of the eigenvalue problem is also given. Chapter 5 is concerned with the solution of the nonlinear inverse problem. There are three major approaches here: the method of successive substitutions, Newton's method, and methods based on minimization of functionals. All of these procedures are iterative in nature and a major problem in the analysis is to find good criteria for the convergence of these iterations. A variety of conditions sufficient for convergence are known, unfortunately these are often inapplicable or impractical in anything but the simplest problems. This difficulty, of course, just reflects our general lack of knowledge about nonlinear problems.

# 3
# NUMERICAL INTEGRATION

As we remarked previously, the solution of the direct problem

$$Tx = y,$$

where $T$ and $x$ are known is, at least from a conceptual viewpoint, relatively simple. Since the main problem here is numerical integration, we will consider only

$$Lx = y, \qquad (3.1)$$

where $L$ is a bounded linear operator $X \to Y$. Nonlinear problems of this type occur less frequently; in any case no essential complication is introduced if the operator is nonlinear. If $Lx$ in (3.1) cannot be evaluated explicitly, it is natural to attempt to replace $x$ with some simpler element $x_n$ such that $Lx_n$ can be evaluated and then compute

$$y_n = Lx_n. \qquad (3.2)$$

Obviously,

$$\|y - y_n\| = \|L(x - x_n)\|, \qquad (3.3)$$

so that

$$\|y - y_n\| \le \|L\| \|x - x_n\|. \qquad (3.4)$$

**THEOREM 3.1.** If $\{x_n\}$ is a sequence of approximations converging to $x$, that is,

$$\lim_{n \to \infty} \|x - x_n\| = 0,$$

then

$$\lim_{n \to \infty} \|y - y_n\| = 0.$$

This is obvious from (3.4) since $L$ is bounded.

The order of convergence of the method can also be determined immediately.

**THEOREM 3.2.** If $x_n$ converges to $x$ with order $p$, then $y_n$ converges to $y$ with at least order $p$.

PROOF. Since we have by definition that

$$\|x - x_n\| \leq Cn^{-p},$$

it follows from (3.4) that

$$\|y - y_n\| \leq C\|L\|n^{-p}. \qquad\blacksquare$$

## 3.1 INTERPOLATORY QUADRATURE RULES

We now consider the specific problem

$$I = Lf = \int_a^b f(t)\,dt. \qquad (3.5)$$

We will take as $X$ the space $C[a,b]$ with maximum norm. Then, quite obviously,

$$\|L\| = |b - a|.$$

Many of the common quadrature rules are derived by replacing $f(t)$ with $f_n(t)$, the piecewise interpolating polynomial on some set of nodes $a \leq t_0 < t_1 < \cdots < t_n \leq b$. Then $I$ is approximated by

$$I_n = \int_a^b f_n(t)\,dt$$

and

$$|I - I_n| \leq |b - a|\,\|f - f_n\|_\infty. \qquad (3.6)$$

**Example 3.1.** The trapezoidal rule

$$\int_a^b f(t)\,dt \cong \frac{b-a}{2}\big[f(a) + f(b)\big]$$

is derived by taking for $f_n$ the linear interpolation polynomial through the endpoints. Then, if $f \in C^{(2)}[a,b]$ and $|f''| \leq M_2$,

$$\|f - f_n\|_\infty \leq \frac{(b-a)^2}{8} M_2$$

from (2.8), and

$$|I - I_n| \leq \frac{(b-a)^3}{8} M_2.$$

The composite trapezoidal rule is constructed by using a piecewise linear polynomial on the uniform partition $t_{i+1} - t_i = h$. Here

$$\|f - f_n\|_\infty \leq \frac{h^2}{8} M_2$$

and

$$|I - I_n| \leq \frac{h^2(b-a)}{8} M_2. \tag{3.7}$$

This error bound is somewhat larger than necessary. By working directly from (3.3) one can show that

$$|I - I_n| \leq \frac{h^2(b-a)}{12} M_2 \tag{3.8}$$

which is the best one can do (Exercise 2). That the error bound (3.4) gives a somewhat worse result should come as no surprise; most of the time a general approach cannot yield as accurate a result as can be obtained by a detailed consideration of a specific case.

**Example 3.2.** If $f(t)$ is not in $C^{(2)}[a,b]$, then the bounds obtained in the previous example no longer hold. Suppose that $f \in C[a,b]$ and that $\|f'\|_\infty \leq M_1$, but $f'$ may be discontinuous at a finite number of points in $[a,b]$. If $f_n$ is the linear interpolation polynomial on the points $a$ and $b$, then a simple geometrical argument shows that

$$\|f - f_n\|_\infty \leq \frac{b-a}{2} M_1,$$

so that

$$|I - I_n| \leq \frac{(b-a)^2}{2} M_1.$$

From this it follows that if the composite trapezoidal rule is applied to functions of this type the convergence is still second order (Exercise 3).

**Example 3.3.** The composite Simpson's rule is derived by introducing nodes $a = t_0 < t_1 < \cdots < t_{2n} = b, t_{i+1} - t_i = h$, and approximating $f(t)$ by a piecewise quadratic polynomial in each subinterval $[t_{2i}, t_{2i+2}]$. For $f \in C^{(3)}[a,b]$, and $|f'''| \leq M_3$, we have from (2.8)

$$\|f - f_n\|_\infty \leq \frac{h^3}{9\sqrt{3}} M_3$$

and

$$|I - I_n| \leq \frac{(b-a)h^3}{9\sqrt{3}} M_3.$$

For $f \in C^{(4)}[a,b]$, $|f^{(4)}| \leq M_4$, it is of course well-known that

$$|I - I_n| \leq \frac{(b-a)h^4}{180} M_4.$$

Again, this improved bound can be established by a more careful analysis; in particular, the fourth-order convergence can be proved by deriving Simpson's rule in a different way (Exercise 5).

**Exercises 3.1**

1. Show by example that Theorem 3.1 is not necessarily true if $L$ is not a bounded operator.
2. Derive the improved error bound (3.8) for the composite trapezoidal rule. Show that this is the best possible result in the sense that there exists a function for which the bound is attained.
3. Let $f(t)$ be a continuous function on $[0,1]$ such that $f''(t)$ exists and is bounded for all $t \in [0,1]$, except at a finite number of points $t_0, t_1, \ldots, t_N$ where $f'(t)$ may have a finite jump discontinuity. Show that if the composite trapezoidal rule is applied to a function of this type, then the error is still $O(h^2)$.
4. Suppose $f \in C^{(2)}$ except at a finite number of points where it may have jump discontinuities. Derive an error bound for the integration of such functions by the composite trapezoidal rule.

5. Derive the composite Simpson's rule by taking for $f_n$ the piecewise third-degree polynomial such that

$$f_n(t_{2i}) = f(t_{2i}), \quad f_n(t_{2i+1}) = f(t_{2i+1}),$$
$$f_n(t_{2i+2}) = f(t_{2i+2}), \quad f_n'(t_{2i+1}) = f'(t_{2i+1}).$$

For $f \in C^{(4)}$ derive an error bound and show that the error is $O(h^4)$.

6. Derive an interpolatory quadrature rule of the form

$$\int_a^b f(t)\, dt \cong w_1 f(a) + w_2 f(b) + w_3 f'(a) + w_4 f'(b)$$

by starting with a simple Hermite interpolating polynomial. For $f \in C^{(4)}$ derive an error bound for this method.

## 3.2   PRODUCT INTEGRATION

In certain cases the direct application of the simple quadrature formulas of the previous section is unsatisfactory. Typically, this happens when $f$ is integrable but not bounded, or if it oscillates rapidly. The technique of product integration can be used to overcome these difficulties. We rewrite the integral as

$$I = \int_a^b p(t)g(t)\, dt, \tag{3.9}$$

where $g(t)$ is well-behaved and $p(t)$ contains the part causing the difficulty. We then approximate $g(t)$ by a simpler function $g_n(t)$ and evaluate

$$I_n = \int_a^b p(t)g_n(t)\, dt. \tag{3.10}$$

For this procedure to be applicable it is necessary that $p(t)$ be simple enough so that the above integral can be evaluated. Fortunately, in most practical situations this is possible and the method is very useful in the integration of ill-behaved functions.

With

$$L[\quad] = \int_a^b p(t)[\quad]\, dt$$

we again have, from (3.4),

$$|I - I_n| \le \|L\|\, \|g - g_n\| \tag{3.11}$$

where, using the maximum norm,

$$\|L\| \leq \int_a^b |p(t)|\, dt.$$

**Example 3.4.** Consider

$$I = \int_0^1 \frac{g(t)}{\sqrt{t}}\, dt$$

where $g(t) \in C^{(2)}[0, 1]$. The product integration analogue of the composite trapezoidal method is derived by approximating $g$ by a piecewise linear function and we get

$$I_n = \sum_{i=0}^n w_i g(t_i).$$

The weights $w_i$ are easily derived (Exercise 1). Also

$$\|L\| = \int_0^1 \frac{dt}{\sqrt{t}} = 2$$

and

$$|I - I_n| \leq \frac{h^2}{4} M_2.$$

Thus the product trapezoidal method is of second order for $g \in C^{(2)}$.

**Example 3.5.** The Fourier integral

$$I = \int_a^b \cos wt \, g(t)\, dt$$

can be evaluated by standard quadrature techniques, but this tends to become inefficient for large $w$ because of the oscillatory behavior of the integrand. The problem can be overcome by the use of product integration.

A particularly simple result is obtained when $a = 2i\pi / w$, $b = 2j\pi / w$, that is, the integration is over one or more periods of the cosine term. Using a quadratic interpolation polynomial we get (Exercise 2)

$$I_n = \frac{2}{(j-i)\pi w} \left[ g(a) - 2g\left(\frac{a+b}{2}\right) + g(b) \right]. \tag{3.12}$$

This technique is known as *Filon's method*. The composite Filon's method can be derived in a straightforward manner, although the weights are somewhat more complicated. Again a simple analysis yields $\|L\| \le |b-a|$ so that the composite Filon's method is at least third order. For a detailed discussion of this see Davis and Rabinowitz (1967, pp. 59–66).

**Exercises 3.2**

1. Derive the weights for the composite product trapezoidal rule in Example 3.4.
2. (a) Derive equation (3.12) starting with a quadratic.
   (b) Derive (3.12) starting with a polynomial of the type used in Exercise 5, Section 3.1.
   (c) Obtain an error bound for Filon's method.
3. (a) Derive the product integration analogue of the composite trapezoidal rule for computing

$$\int_0^1 \log_e tg(t)\,dt.$$

   (b) Derive an error bound for this method if $g \in C^{(2)}[0,1]$.

**3.3 ANOTHER APPROACH TO NUMERICAL INTEGRATION**

Whatever motivation we use for deriving numerical integration rules, the approximation formulas always have the form

$$I_n = \sum_{i=1}^{n} w_i f(t_i).$$

Instead of considering this as derived by applying $L$ to some element $x_n$ we can also consider it as applying the summation operator directly to $x$, that is we can consider numerical integration formulas as resulting from the application of a sequence of operators $L_n$, approximating $L$ in some sense, and such that

$$y_n = L_n x \qquad (3.13)$$

can be computed. Then

$$\|y - y_n\| = \|(L - L_n)x\| \le \|L - L_n\|\,\|x\|. \qquad (3.14)$$

If we could claim that

$$\lim_{n \to \infty} \|L - L_n\| = 0,$$

then (3.14) would establish the convergence of such schemes. Unfortunately, one cannot make such a claim, as is shown by the following simple argument. Let

$$Lf = \int_0^1 f(t)\,dt$$

and consider the Riemann sums

$$L_n f = h \sum_{i=0}^{n-1} f(ih), \qquad h = 1/n.$$

Then considering $L$ and $L_n$ as operators on $C[0,1]$ with maximum norm,

$$\lim_{n \to \infty} \|L - L_n\| \not\to 0. \tag{3.15}$$

To see this, fix $n$ and let $\varphi_n$ be the continuous, piecewise linear function

$$\varphi_n(ih) = 0, \qquad i = 0, 1, \ldots, n$$
$$\varphi_n((i+1/2)h) = 1.$$

Then $L\varphi_n = \frac{1}{2}$ and $L_n\varphi_n = 0$, so that

$$\|L - L_n\| = \sup_{\|f\|=1} \|(L - L_n)f\| \geq \frac{1}{2}$$

for all $n$. To make any claim about the convergence of (3.13) we must take a closer look at the problem.

**THEOREM 3.3.** Let $\varphi_1, \varphi_2, \ldots$ be a basis for $X$ and let $X_n = \mathrm{span}$ $(\varphi_1, \varphi_2, \ldots, \varphi_n)$. Let $L_n$ be a sequence of uniformly bounded linear operators with

$$\sup_n \|L_n\| \leq l,$$

such that $L_n$ is exact on $X_n$, that is,

$$L_n x_n = L x_n \tag{3.16}$$

for all $x_n \in X_n$. Then

$$y_n = L_n x \rightarrow Lx = y \qquad (3.17)$$

and

$$\|y - y_n\| \leq (\|L\| + l)\epsilon_n, \qquad (3.18)$$

where

$$\epsilon_n = \inf_{x_n \in X_n} \|x - x_n\|.$$

PROOF. For every $x_n \in X_n$

$$\|y - y_n\| = \|(L - L_n)x\| = \|Lx - Lx_n + Lx_n - L_n x_n + L_n x_n - L_n x\|$$
$$\leq \|L(x - x_n)\| + \|L_n(x - x_n)\|$$

since $Lx_n - L_n x_n = 0$. Thus,

$$\|y - y_n\| \leq (\|L\| + l)\|x - x_n\|$$

for all $x_n \in X_n$ and (3.18) follows. Since the $\varphi_i$ are a basis for $X$ there exists a sequence of $x_n \in X_n$ such that

$$\lim_{n \to \infty} \|x - x_n\| = 0$$

and (3.17) follows. ∎

**Example 3.6.** In the Gauss–Legendre quadrature method

$$I = Lf = \int_{-1}^{1} f(t)\,dt$$

is approximated by

$$I_n = L_n f = \sum_{i=1}^{n} w_i f(t_i),$$

where the $t_i$ are the roots of the $n$th Legendre polynomial and the $w_i$ the appropriate weights. It is known (Davis and Rabinowitz, 1967, Sec. 2.7) that 1) the $n$-point Gauss–Legendre quadrature is exact for polynomials up to degree $2n-1$ and 2) the weights are all positive. Since the quadrature is exact for $f = 1$ we have

$$\sum_{i=1}^{n} w_i = \sum_{i=1}^{n} |w_i| = 2$$

and hence $\|L_n\| = 2$. Since by the Weierstrass theorem the powers of $t$ are a basis for $C[-1, 1]$, Theorem 3.3 gives the well-known result that the Gauss–Legendre quadrature rules converge for all continuous integrands.

**Exercises 3.3**

1. Show that for the $n$-point Gauss–Legendre rule with $f \in C^{(2n)}[-1, 1]$

$$|I - I_n| \leq \frac{1}{2^{2n-3}(2n)!} M_{2n},$$

   where $M_{2n} = \max_{-1 \leq t \leq 1} |f^{(2n)}(t)|$.
2. With the Gauss–Chebyshev rule

$$I = \int_{-1}^{1} \frac{f(t)}{\sqrt{1 - t^2}} dt$$

   is approximated by

$$I_n = \frac{\pi}{n} \sum_{i=1}^{n} f(t_i),$$

   where the $t_i$ are the roots of $T_n(t)$. For $f \in C^{(2n)}[-1, 1]$ show that

$$|I - I_n| \leq \frac{\pi}{2^{2n-2}(2n)!} M_{2n}.$$

# 4

# THE APPROXIMATE
# SOLUTION OF LINEAR
# OPERATOR EQUATIONS

We begin our study of the inverse problem by considering the linear case

$$Lx = y. \tag{4.1}$$

Many physical situations lead directly to linear equations, while nonlinear problems are often "linearized" as we see in Chapter 5. Hence numerical techniques for the linear inverse problem are of prime importance and both practical and theoretical questions have received considerable attention.

It has already been remarked that if $L$ has an inverse, then formally (4.1) can be reduced to a direct problem, $x = L^{-1}y$. This, however, is of little use as $L^{-1}$ is rarely known explicitly. Nevertheless, some information about $L^{-1}$, for example, whether it exists and is bounded, is generally desirable. Before starting any numerical computation it is advisable to (and often dangerous not to) begin by considering such theoretical questions as the existence and uniqueness of the solution and its sensitivity to small perturbations.

Hadamard introduced the notion of a well-posed problem (Isaacson and Keller, 1966, pp. 21–23). Loosely speaking, (4.1) is said to be well-posed if it has a unique solution which depends continuously on the parameters of the problem (e.g., coefficients in the operator, boundary values, etc.). According to this definition a problem is either well-posed or it is not, but in numerical work this distinction is not as clearcut. As anyone with experience in matrix computation knows, an effective criterion for deciding whether a system of linear equations is well-posed is not easy to come by. Even if one starts with a singular system, round-off error quickly destroys the singularity, and numerically one can "invert" a singular matrix. The answer of course is meaningless. On the other hand, it is equally unrewarding to attempt to compute the inverse of a $30 \times 30$ Hilbert matrix, although strictly speaking that matrix is nonsingular.

We are thus primarily interested in the sensitivity of the solution to perturbation in the parameters. A *condition number* is a measure of this sensitivity. For example, let $L$ in (4.1) be an $n \times n$ matrix with elements $a_{ij}$ and let $x$ and $y$ be vectors with components $x_i$ and $y_i$, respectively. Then a complete set of condition numbers is given by

$$\frac{\partial x_i}{\partial a_{jk}}, \frac{\partial x_i}{\partial y_j}, \qquad i,j,k = 1,2,\ldots,n.$$

As there are $n^2(n+1)$ of these, the definition does not appear very useful. By sacrificing some detail we can arrive at a more manageable concept. Assume that $L$ is perturbed by $\Delta L$ and $y$ by $\Delta y$, giving a perturbed solution $x + \Delta x$. Then

$$(L+\Delta L)(x+\Delta x) = y + \Delta y. \qquad (4.2)$$

We now look for a relation between $\|\Delta x\|$ and the perturbations $\|\Delta y\|$ and $\|\Delta L\|$.

**THEOREM 4.1.** If $L:X\to Y$, where $X$ and $Y$ are Banach spaces, has a bounded inverse on $Y$ and if the perturbation $\Delta L:X\to Y$ satisfies

$$\|\Delta L\| < 1/\|L^{-1}\|, \qquad (4.3)$$

then

$$\|\Delta x\| \leq \frac{\|L^{-1}\|}{1-\|L^{-1}\|\,\|\Delta L\|}(\|\Delta y\| + \|\Delta L\|\,\|x\|), \qquad (4.4)$$

and

$$\frac{\|\Delta x\|}{\|x\|} \leq \frac{k}{1-k\|\Delta L\|/\|L\|}\left[\frac{\|\Delta y\|}{\|y\|} + \frac{\|\Delta L\|}{\|L\|}\right], \qquad (4.5)$$

where $k = \|L\|\,\|L^{-1}\|$.

PROOF. Anticipating a result to be proved in Theorem 4.5, we know that if $L$ has a bounded inverse, and if (4.3) holds, then $L + \Delta L$ has an inverse and

$$\|(L+\Delta L)^{-1}\| \leq \frac{\|L^{-1}\|}{1-\|L^{-1}\|\,\|\Delta L\|}.$$

Subtracting (4.1) from (4.2)

$$(L+\Delta L)\Delta x = \Delta y - \Delta Lx$$

and

$$\|\Delta x\| \leq \|(L+\Delta L)^{-1}\|(\|\Delta y\| + \|\Delta Lx\|)$$

$$\leq \frac{\|L^{-1}\|}{1-\|L^{-1}\|\,\|\Delta L\|}(\|\Delta y\| + \|\Delta L\|\,\|x\|).$$

Also, $\|x\| \geq \|y\|/\|L\|$, so that

$$\frac{\|\Delta x\|}{\|x\|} \leq \frac{\|L^{-1}\|}{1-\|L^{-1}\|\,\|\Delta L\|}\left[\frac{\|\Delta y\|\,\|L\|}{\|y\|} + \|\Delta L\|\right].$$

from which (4.5) follows. ∎

Thus, the sensitivity of the solution depends primarily on the magnitude of $\|L^{-1}\|$; if the relative sensitivity is to be determined, then $\|L\|\,\|L^{-1}\|$ is a good measure for it. It is customary to call

$$k = \|L\|\,\|L^{-1}\| \tag{4.6}$$

the (relative) condition number for (4.1). It is also customary, although somewhat vague, to say that a problem is ill-conditioned if $k$ is large, and well-conditioned if $k$ is of order unity. Thus, in numerical analysis the strict mathematical distinction between well-posed and not well-posed problems translates into the somewhat less precise concept of well-conditioned and ill-conditioned problems.

## 4.1 SOME THEOREMS ON LINEAR OPERATORS

By definition, (4.1) has a solution if $y \in \Re(L)$. The solution is unique if the mapping is one-to-one, that is, if the homogeneous equation has only a trivial solution.

**DEFINITION 4.1.**    The set of all $x$ such that

$$Lx = 0$$

is called the *null-space* of $L$ and denoted by $\mathfrak{N}(L)$.

**THEOREM 4.2.**    The inverse of $L$ exists [on $\mathfrak{R}(L)$] if and only if $\mathfrak{N}(L)$ contains only the zero element.

**PROOF.**

(a) Assume that $\mathfrak{N}(L)$ contains only 0. Then there exists no $z \neq 0$ such that $Lz = 0$. This implies that $L$ is one-to-one, since otherwise we could find $x_1$ and $x_2$ such that $Lx_1 = Lx_2$ or $L(x_1 - x_2) = 0$, which is a contradiction. Hence $L$ is one-to-one and onto $\mathfrak{R}(L)$ so that $L^{-1}$ exists.

(b) Assume that there is a $z \neq 0$ in $\mathfrak{N}(L)$. Then if $Lx = y$ we also have $L(x + \alpha z) = y$. Hence $L$ is not one-to-one and $L^{-1}$ does not exist. ∎

In this chapter we concern ourselves mainly with the problem of solving (4.1) under the assumption that $L$ has a bounded inverse. In Section 4.5 we briefly discuss the case when $L$ does not have an inverse by considering some approximate methods for computing eigenvalues.

We begin by developing some important theorems on the existence and the representation of the inverses of certain linear operators.

**THEOREM 4.3.**    Let $X$ and $Y$ be Banach spaces. Then $L[X, Y]$ with norm defined by (1.28) is also a Banach space.

**PROOF.**    Let $\{L_m\}$ be a Cauchy sequence in $L[X, Y]$ and consider the sequence $y_m = L_m x$ for a given $x \in X$. Since

$$\|y_{m+p} - y_m\| \leq \|L_{m+p} - L_m\| \, \|x\|,$$

the sequence $\{y_m\}$ is a Cauchy sequence and therefore has a limit $y \in Y$. Thus, with each $x \in X$ there is associated a unique $y \in Y$ through this limit. Let $L$ represent this mapping. It is obvious that $L$ is linear; to show that it is bounded consider

$$\|Lx_1 - Lx_0\| \leq \|Lx_1 - L_m x_1\| + \|L_m x_0 - Lx_0\| + \|L_m(x_1 - x_0)\|.$$

Since $\{L_m\}$ is a Cauchy sequence, we must have $\|L_m\| \leq K$, for sufficiently large $m$. Also, for sufficiently large $m$, we have for any $c > 0$

$$\|L_m x_0 - Lx_0\| \leq c\|x_1 - x_0\|,$$

and

$$\|L_m x_1 - L x_1\| \le c\|x_1 - x_0\|.$$

Thus,

$$\|L x_1 - L x_0\| \le (2c + K)\|x_1 - x_0\|$$

and $L$ is bounded.

Finally, since $\{L_m\}$ is a Cauchy sequence we have, for $m, p$ sufficiently large

$$\|L_{m+p} x - L_m x\| \le \epsilon\|x\|, \qquad \text{for all } x \in X.$$

Letting $p$ go to infinity gives

$$\|L x - L_m x\| \le \epsilon\|x\|, \qquad \text{for all } x \in X.$$

Therefore $L_m$ converges to $L$ in norm and the Cauchy sequence $\{L_m\}$ has a limit $L \in L[X, Y]$ so that $L[X, Y]$ is a Banach space. ∎

In classical analysis, operators are often analyzed by considering them to be perturbations of simpler operators. This approach is particularly useful in numerical analysis, since the numerical solution generally requires that we replace $L$ by a more amenable "neighboring" operator. The next several theorems deal with the relation between the inverses of such neighboring operators.

**THEOREM 4.4.** A linear operator $L$ from a Banach space $X$ onto a Banach space $Y$ has a bounded inverse if and only if there exists some bounded linear operator $K: Y \to X$ such that $K^{-1}$ exists and

$$\|I - KL\| < 1. \tag{4.7}$$

PROOF. The necessity is obvious, since if $L^{-1}$ exists we can put $K = L^{-1}$. Consider now the sequence of operators

$$Q_m = \sum_{n=0}^{m} (I - KL)^n K.$$

$\{Q_m\}$ is a Cauchy sequence in $L[Y, X]$ so that by the previous theorem it has a limit

$$Q = \sum_{n=0}^{\infty} (I - KL)^n K$$

in $L[Y,X]$, and

$$\|Q\| \leq \sum_{n=0}^{\infty} \|(I - KL)^n\| \|K\|$$

$$\leq \|K\| \sum_{n=0}^{\infty} \|I - KL\|^n$$

$$\leq \frac{\|K\|}{1 - \|I - KL\|}. \qquad (4.8)$$

For a given $y \in Y$ let $x$ be defined by

$$x = Qy = \sum_{n=0}^{\infty} (I - KL)^n Ky.$$

Multiplying by $(I - KL)$ we have

$$(I - KL)x = \sum_{n=1}^{\infty} (I - KL)^n Ky = x - Ky,$$

$$KLx = Ky,$$

$$Lx = y.$$

Thus, the equation $Lx = y$ has at least one solution $x = Qy$. If there were another solution $x^* \neq x$, then

$$L(x - x^*) = 0,$$

$$(I - KL)(x - x^*) = x - x^*,$$

$$\|I - KL\| \|x - x^*\| \geq \|x - x^*\|,$$

implying that

$$\|I - KL\| \geq 1,$$

which contradicts assumption (4.7). Thus, $Lx = y$ has a unique solution for all $y$ and $L^{-1}$ exists. It also follows that $Q$ is a representation for the inverse of $L$ and we can write

$$L^{-1} = \sum_{n=0}^{\infty} (I - KL)^n K, \qquad (4.9)$$

with the bound

$$\|L^{-1}\| \leq \frac{\|K\|}{1 - \|I - KL\|}. \qquad (4.10)$$

The series (4.9) is the *Neumann Expansion* for $L^{-1}$.     ■

To exhibit more clearly the relation between the inverse of $L$ and the inverse of a neighboring operator we rewrite the above theorem in a somewhat different form.

**THEOREM 4.5.** Let $L$ be a linear operator between two Banach spaces $X$ and $Y$, and let $\Delta L \in L[X, Y]$ satisfy

$$\|\Delta L\| < \frac{1}{\|L^{-1}\|}.$$

Then $(L + \Delta L)$ has a bounded inverse given by

$$(L + \Delta L)^{-1} = \sum_{n=0}^{\infty} (L^{-1}\Delta L)^n L^{-1},$$

with

$$\|(L + \Delta L)^{-1}\| \leq \frac{\|L^{-1}\|}{1 - \|L^{-1}\| \|\Delta L\|}. \tag{4.11}$$

PROOF.  Put $L = L + \Delta L$ and $K = L^{-1}$ in Theorem 4.4.  ∎

A particularly simple version of these results is obtained if we let $L = I - P$ and set $K = I$. Then if

$$\|P\| \leq k < 1,$$

we have from (4.9)

$$(I - P)^{-1} = \sum_{n=0}^{\infty} P^n$$

and

$$\|(I - P)^{-1}\| \leq \frac{1}{1 - k}. \tag{4.12}$$

The solution of the equation

$$(I - P)x = y$$

is given by

$$x = \sum_{n=0}^{\infty} P^n y. \tag{4.13}$$

Usually it is not convenient to compute the powers of $P$ explicitly; one uses instead a simple, equivalent iteration process.

**THEOREM 4.6.**   Consider the sequence

$$x_{m+1} = Px_m + y \qquad (4.14)$$

with $x_0 = y$. Then (4.13) truncated at the $m$th term is equal to $x_m$, that is,

$$x_m = \sum_{n=0}^{m} P^n y, \qquad (4.15)$$

and

$$\lim_{m \to \infty} x_m = x, \qquad (4.16)$$

where $x$ is the solution of $(I - P)x = y$.

PROOF.   By induction

$$x_{m+1} = P \sum_{n=0}^{m} P^n y + y$$

$$= \sum_{n=0}^{m+1} P^n y.$$

Since the relation is obviously true, for $m = 1$ (4.15) and (4.16) follow.   ∎

The iteration procedure (4.14) is the *Picard iteration method* or the *method of successive substitutions*.

The next theorem is frequently useful in obtaining a bound on the inverse of an operator.

**THEOREM 4.7.**   Let $L$ be a linear operator $X \to Y$. If there exists an $m > 0$ such that

$$\|Lx\| \geq m\|x\|, \qquad \text{for all } x \in X, \qquad (4.17)$$

then $L^{-1}$ exists on $\mathcal{R}(L)$ and

$$\|L^{-1}\| \leq 1/m. \qquad (4.18)$$

PROOF.   $L$ must be one-to-one since $\|L(x_1 - x_2)\| \geq m\|x_1 - x_2\|$, so that $Lx_1 = Lx_2$ implies that $x_1 = x_2$. Thus, for every $y \in \mathcal{R}(L)$ there exists a unique $x$ such that $Lx = y$ and

$$\|y\| = \|Lx\| \geq m\|x\| = m\|L^{-1}y\|.$$

Therefore,

$$\|L^{-1}y\|/\|y\| \leq 1/m, \qquad \text{for all } y,$$

and hence by the definition of the operator norm, (4.18) follows. ∎

Once an approximate solution of (4.1) has been computed, its accuracy can be estimated if a bound on $\|L^{-1}\|$ is available.

**THEOREM 4.8.** Let $\hat{x} \in X$ be an approximate solution of (4.1) such that

$$L\hat{x} - y = r. \tag{4.19}$$

Then

$$\|\hat{x} - x\| \leq \|L^{-1}\| \, \|r\|. \tag{4.20}$$

PROOF. Subtracting (4.1) from (4.19) we have

$$L(\hat{x} - x) = r,$$
$$\hat{x} - x = L^{-1}r,$$

and (4.20) follows. ∎

$r$ is the *residual* of the approximate solution. We expect that a small residual indicates a good approximation; however, when $\|L^{-1}\|$ is large, this is not necessarily the case.

**Example 4.1.** Consider the approximate solution of the matrix system

$$\begin{bmatrix} 1 & 0.1 & 0.01 & 0.001 \\ 0.1 & 1 & 0.01 & 0.001 \\ 0.1 & 0.01 & 1 & 0.001 \\ 0.1 & 0.01 & 0.001 & 1 \end{bmatrix} \begin{bmatrix} x_1 \\ x_2 \\ x_3 \\ x_4 \end{bmatrix} = \begin{bmatrix} 1 \\ 1 \\ 1 \\ 1 \end{bmatrix}.$$

We will use the vector norm $\|x\| = \max|x_i|$, so that the induced matrix norm is the maximum row-sum norm (Example 1.10). Writing the matrix as $L = I - P$ we have $\|P\| = 0.111$ and from (4.12) the system has a unique solution with

$$\|L^{-1}\| \le \frac{1}{1-0.111} < 1.2.$$

The successive approximations computed by (4.14) are

$$\mathbf{x}_1 = \begin{pmatrix} 0.88900 \\ 0.88900 \\ 0.88900 \\ 0.88900 \end{pmatrix}, \mathbf{x}_2 = \begin{pmatrix} 0.90132 \\ 0.90132 \\ 0.90132 \\ 0.90132 \end{pmatrix}, \mathbf{x}_3 = \begin{pmatrix} 0.89995 \\ 0.89995 \\ 0.89995 \\ 0.89995 \end{pmatrix}.$$

The residual for $\mathbf{x}_3$ is

$$\mathbf{r} = L\mathbf{x}_3 - \begin{pmatrix} 1 \\ 1 \\ 1 \\ 1 \end{pmatrix} = - \begin{pmatrix} 0.00015 \\ 0.00015 \\ 0.00015 \\ 0.00015 \end{pmatrix}$$

and hence $\|\mathbf{x}_3 - \mathbf{x}\| \le 1.8 \times 10^{-4}$.

**Example 4.2.**   Consider the differential equation

$$Dx = \frac{dx}{dt} - x = e^t \tag{4.21}$$

$$x(0) = 0$$

in the interval $[0, 1/2]$. It is a standard result in the theory of ordinary differential equations that equations of this type have a unique solution under rather general conditions, certainly when the right-hand side is a continuous function. Thus, $D : C^{(1)} \to C$ has an inverse. To find a bound on $\|D^{-1}\|$ replace the right-hand side of (4.21) with an arbitrary function $f$ and integrate. Then

$$x(t) = \int_0^t f(s)\,ds + \int_0^t x(s)\,ds. \tag{4.22}$$

Using the maximum norm

$$\|x\| \le \frac{1}{2}\|f\| + \frac{1}{2}\|x\|,$$

$$\|f\| \ge \|x\|.$$

Since

$$\|Dx\| = \|f\| \ge \|x\|$$

we have from Theorem 4.7 that $\|D^{-1}\| \leq 1$. If we now take a trial solution obtained from a power series expansion of (4.21),

$$\hat{x} = t + t^2 + t^3/2,$$

then

$$r = D\hat{x} - e^t = -2t^3/3 - t^4/24 + \cdots$$
$$\|r\| < 1/10,$$

so that

$$\|\hat{x} - x\| < 1/10.$$

Several points are worth noticing in this example.

1. The actual error in the approximate solution is about 0.01, so that the bound is somewhat pessimistic. This is not unusual in this type of analysis.

2. With the chosen norm $D$ is not bounded, but Theorems 4.7 and 4.8 are still applicable. The Picard iteration, on the other hand, cannot be used directly, but a reformulation of the original equation in terms of the integral equation (4.22) makes it possible to compute successive approximations by (4.14).

3. In our treatment we ignored the initial condition. This causes no difficulty here; effectively we consider $X$ to be the set of functions $C^{(1)}[0, 1/2]$ with $x(0)=0$, which is a linear space. When we have nonzero initial or boundary conditions somewhat more care has to be taken.

### Exercises 4.1

1. For the $2 \times 2$ linear system

$$a_{11}x_1 + a_{12}x_2 = y_1$$
$$a_{21}x_1 + a_{22}x_2 = y_2$$

compute the condition numbers $\partial x_1/\partial a_{11}$, $\partial x_1/\partial y_1$, and $k = \|A\|\|A^{-1}\|$.

2. Verify that the $Q_m$ in Theorem 4.4 form a Cauchy sequence.

3. If $x_m$ is computed by the Picard iteration (4.14), show that for $\|P\| = k < 1$,

$$\|x_m - x\| \leq \frac{k^{m+1}}{1-k}\|y\|.$$

4. Find an approximate solution of the system

$$x_1 + 0.1x_2 - 0.1x_3 + 0.1x_4 = 1$$
$$0.1x_1 + x_2 + 0.1x_3 - 0.1x_4 = 2$$
$$0.1x_x + 0.1x_2 + x_3 - 0.1x_4 = 0$$
$$0.1x_1 + 0.1x_2 + 0.1x_3 + x_4 = 1$$

by applying the method of successive substitutions four times. Find a bound for the error in the result.

5. Show that the integral equation

$$x(t) + 0.1 \int_0^1 e^{ts}x(s)\,ds = 1, \qquad 0 \le t \le 1,$$

has a unique continuous solution. Find an approximate solution accurate to 0.1.

6. Find an approximate solution of

$$x' - x = 0$$
$$x(0) = 1, \qquad 0 \le t \le \frac{1}{2}$$

by power series expansion up to $t^3$. Estimate the accuracy of the result. Make sure your argument takes the nonhomogeneous initial condition into account.

7. Let $L$ be a bounded linear operator $X \to X$ with a bounded inverse. Show that the equation

$$(L + \alpha I)z = y$$

has a unique solution for sufficiently small $|\alpha|$. If $Lx = y$ find a bound for $\|z - x\|$ in terms of $|\alpha|$, $\|L\|$, $\|L^{-1}\|$, and $\|y\|$.

8. In Theorem 4.5 the condition that $\|\Delta L\|$ be small can be replaced by the less restrictive condition that $\|L^{-1}\Delta L\|$ be small. Restate the theorem in terms of this condition and prove it.

9. Prove that a diagonally dominant matrix is nonsingular.

## 4.2   APPROXIMATE EXPANSION METHODS

We have seen in the last section that, at least for well-conditioned problems, a small residual implies a good approximation. This immediately suggests that one might try to construct approximation methods so as to make the residual small (in some sense). To be more specific, consider a sequence of elements $\varphi_{ni} \in X, i = 1, 2, \ldots, n, n = 1, 2, \ldots$ . We will assume that the set $\{\varphi_{ni}\}$ is closed in the normed linear space $X$ and use $\Phi_n$ to denote the span of $\{\varphi_{n1}, \varphi_{n2}, \ldots, \varphi_{nn}\}$. We now look for an approximation $x_n \in \Phi_n$ of the form

$$x_n = x_n(\boldsymbol{\alpha}) = \sum_{i=1}^{n} \alpha_i \varphi_{ni}, \tag{4.23}$$

where the coefficients $\alpha_i$ are to be chosen according to some appropriate criterion. If we write

$$r_n(\boldsymbol{\alpha}) = Lx_n(\boldsymbol{\alpha}) - y,$$

then it follows from (4.20) that, if $L$ has a bounded inverse and if $\boldsymbol{\alpha}$ is chosen so that

$$\lim_{n \to \infty} \|r_n(\boldsymbol{\alpha})\| = 0,$$

then $x_n$ converges to $x$. Since by assumption the set $\{\varphi_{ni}\}$ is closed in $X$ there exists an expansion of $x$ (generally containing an infinite number of terms) in terms of $\{\varphi_{ni}\}$. The $x_n$ in (4.23) can then be considered as an approximation to this expansion, hence we use the term *approximate expansion methods* to denote this type of approach. As we shall see, a number of well-known numerical methods fall into this category; they differ primarily in the way in which the coefficients $\alpha_i$ are selected.

Since the closeness of the approximation is determined by the smallness of the residual, it is natural to choose the $\alpha_i$ such that this residual is minimized. Thus, we can take $\boldsymbol{\alpha} \in R^n$ such that

$$\|r_n(\boldsymbol{\alpha})\| \le \|r_n(\boldsymbol{\beta})\|, \qquad \text{for all } \boldsymbol{\beta} \in R^n. \tag{4.24}$$

We will leave it to the reader to show that such a minimizing $\boldsymbol{\alpha}$ exists. Methods of this type are referred to as *minimum residual methods*.

**THEOREM 4.9.** If $L$ is a bounded linear operator with a bounded inverse, then $x_n$ computed by the minimum residual method converge to the exact solution $x$.

**PROOF.** Since the $\varphi_{ni}$ are closed in $X$ there exists, for all sufficiently large $n$, an approximation $x_n(\boldsymbol{\beta})$ such that

$$\|x_n(\boldsymbol{\beta}) - x\| \le \epsilon.$$

Now

$$r_n(\beta) = Lx_n(\beta) - y = Lx_n(\beta) - Lx,$$
$$\|r_n(\beta)\| \leq \|L\|\epsilon.$$

For the minimizing solution $x_n(\alpha)$

$$\|r_n(\alpha)\| \leq \|r_n(\beta)\|,$$

so that

$$\|x_n(\alpha) - x\| \leq \|L^{-1}\| \|r_n(\alpha)\|$$
$$\leq \|L^{-1}\| \|r_n(\beta)\|$$
$$\leq \|L^{-1}\| \|L\|\epsilon.$$

Since $\epsilon$ can be made arbitrarily small, convergence follows.    ∎

The assumption that $L$ is bounded is restrictive and in fact not necessary. All that is really needed is that there exists a $\beta$ such that $\|r_n(\beta)\|$ can be made arbitrarily small. This is often true even if $L$ is not bounded and the preceding theorem can be generalized to these cases (see Exercise 5).

Perhaps the best-known minimum residual method is the *method of least squares*. If $X$ and $Y$ are inner product spaces, then

$$\|r_n(\alpha)\|^2 = (r_n, r_n) = (Lx_n - y, Lx_n - y)$$
$$= \sum_{i=1}^{n} \sum_{j=1}^{n} \alpha_i \alpha_j (L\varphi_{ni}, L\varphi_{nj}) - 2 \sum_{i=1}^{n} \alpha_i (L\varphi_{ni}, y) + (y, y).$$

Differentiating with respect to $\alpha_k$, $k = 1, 2, \ldots, n$ and setting the result to zero we get the *generalized normal equations*.

$$A\alpha = b, \tag{4.25}$$

where $A$ is a matrix with elements

$$a_{ij} = (L\varphi_{ni}, L\varphi_{nj}), \tag{4.26}$$

and $b$ is a vector with components

$$b_i = (L\varphi_{ni}, y). \tag{4.27}$$

**Example 4.3.** Use the least squares method to find an approximate solution of the equation

$$Lx = x'' - x = 1, \qquad 0 \le t \le 1,$$
$$x(0) = x(1) = 0.$$

For simplicity we choose expansion functions that satisfy the boundary conditions, for example,

$$\varphi_{21} = \sin \pi t,$$
$$\varphi_{22} = \sin 2\pi t.$$

Then

$$L\varphi_{21} = -(\pi^2 + 1)\sin \pi t$$
$$L\varphi_{22} = -(4\pi^2 + 1)\sin 2\pi t.$$

With the inner product defined by

$$(f,g) = \int_0^1 f(t)g(t)\,dt$$

the normal equations are

$$\begin{bmatrix} \dfrac{(\pi^2 + 1)^2}{2} & 0 \\[3mm] 0 & \dfrac{(4\pi^2 + 1)^2}{2} \end{bmatrix} \begin{pmatrix} \alpha_1 \\ \alpha_2 \end{pmatrix} = \begin{bmatrix} \dfrac{-2(\pi^2 + 1)}{\pi} \\[3mm] 0 \end{bmatrix},$$

and the approximate solution is

$$x_2 = -\frac{4}{\pi(\pi^2 + 1)}\sin \pi t.$$

This agrees with the exact solution

$$x = \frac{(e^t + e^{1-t})}{(e + 1)} - 1$$

to within 0.006.

**Example 4.4.**   Find an approximate solution to

$$x'' - x = 1, \qquad 0 \le t \le 1,$$
$$x(0) = \alpha, \qquad x(1) = \beta.$$

In applying the general theory here we must be careful since it is not possible to find trial functions which satisfy the inhomogeneous boundary conditions and at the same time constitute a basis for a linear space. There are several ways around this difficulty.

(a)   We can transform the problem so that the boundary conditions are homogeneous. This is simple in the present case. Introduce $z(t)$ by

$$x(t) = z(t) + (\beta - \alpha)t + \alpha,$$

then $z(t)$ satisfies

$$z'' - z = 1 + (\beta - \alpha)t + \alpha,$$
$$z(0) = z(1) = 0.$$

and we can now use the technique of the previous example.

(b)   The problem can be treated directly by using the notion of a product space introduced in Definition 1.21. Let $X = C^{(2)}[0, 1]$, $Y = C[0, 1] \times R^2$ and define $L : X \rightarrow Y$ by

$$Lx = (x'' - x, x(0), x(1)).$$

The original equation can now be written as

$$Lx = (1, \alpha, \beta).$$

On the space $Y$ we now introduce an inner product defined by

$$((f, c_1, c_2), (g, c_3, c_4)) = \int_0^1 w(t) f(t) g(t)\, dt + w_1 c_1 c_3 + w_2 c_2 c_4,$$

with $w(t) > 0$, $w_1 > 0$, $w_2 > 0$. The solution of the normal equations in this setting yields the answer $x_n$ which minimizes the residual

$$\int_0^1 w(t)(1 - x_n'' + x_n)^2\, dt + w_1(x_n(0) - \alpha)^2 + w_2(x_n(1) - \beta)^2.$$

Now neither the equation nor the boundary conditions are satisfied exactly. How well they are approximated depends on the relative magnitudes of $w(t), w_1$, and $w_2$.

The least squares method minimizes the residual in the 2-norm; other natural choices would be the 1-norm or the maximum norm, but these lead to much more complicated computational problems and have therefore been used less frequently. While it is intuitively attractive to choose $\alpha$ such that the residual is minimized it is of course possible to use some other criterion. For instance, in a Hilbert space setting we can require that $r_n$ be orthogonal to the subspace $\Phi_n$. In this case we must have

$$(r_n, \varphi_{ni}) = (Lx_n - y, \varphi_{ni}) = 0, \qquad i = 1, 2, \cdots, n,$$

leading to the linear system

$$A\alpha = b, \tag{4.28}$$

where

$$a_{ij} = (\varphi_{ni}, L\varphi_{nj}), \tag{4.29}$$

$$b_i = (\varphi_{ni}, y). \tag{4.30}$$

This technique is known as *Galerkin's method*.* Since this method does not minimize the residual, Theorem 4.9 is not applicable. However, under certain simple assumptions convergence can be established without difficulty.

**DEFINITION 4.2.** A linear operator $L$ with domain and range in a Hilbert space is said to be *positive-definite* if there exists a $\gamma > 0$ such that

$$(Lx, x) \geq \gamma \|x\|^2 \tag{4.31}$$

for all $x$.

**THEOREM 4.10.** Let $L$ be a positive-definite bounded linear operator from a Hilbert space into itself. Then the $x_n$ computed by the Galerkin method converge to the true solution, that is,

$$\|x_n - x\| \to 0 \qquad \text{as } n \to \infty.$$

**PROOF.** Again, since the $\varphi_{ni}$ are a basis for $X$ there exists, for sufficiently large $n$, a $\varphi \in \Phi_n$ such that $\|x - \varphi\| \leq \epsilon$. Now for all $\psi \in \Phi_n$

$$(L(\varphi - x), \psi) = (L(\varphi - x_n), \psi) + (L(x_n - x), \psi)$$
$$= (L(\varphi - x_n), \psi) + (Lx_n - y, \psi)$$
$$= (L(\varphi - x_n), \psi).$$

*This method and some of its variations are also known under the names Bubnov–Galerkin, Galerkin–Petrov, Ritz–Galerkin, or Rayleigh–Ritz methods.

Setting $\psi = \varphi - x_n$ we have

$$|(L(\varphi - x_n), \varphi - x_n)| = |(L(\varphi - x), \varphi - x_n)|$$
$$\leq \|L\| \, \|\varphi - x_n\| \epsilon.$$

Since $L$ is positive definite

$$\|\varphi - x_n\|^2 \leq \frac{1}{\gamma}(L(\varphi - x_n), \varphi - x_n) \leq \frac{1}{\gamma}\|L\| \, \|\varphi - x_n\| \epsilon,$$

$$\|\varphi - x_n\| \leq \frac{1}{\gamma}\|L\|\epsilon,$$

and

$$\|x - x_n\| \leq \|x - \varphi\| + \|\varphi - x_n\|$$
$$\leq (1 + \|L\|/\gamma)\epsilon. \qquad\blacksquare$$

As in the method of least squares the restriction that $L$ be bounded can be removed under certain conditions (Exercise 5).

Both the Galerkin method and the method of least squares can be considered special cases of a more general formulation. Let $L_i$, $i = 1, 2, \ldots, n$ be a set of linear functionals and select $\alpha$ such that

$$L_i L x_n = L_i \sum_{j=1}^{n} \alpha_j L\varphi_{nj} = L_i y. \tag{4.32}$$

Thus the $\alpha_i$ are determined by the linear system $A\alpha = \mathbf{b}$ with

$$a_{ij} = L_i(L\varphi_{nj}) \tag{4.33}$$

and

$$b_i = L_i y. \tag{4.34}$$

With $L_i[\ ] = (L\varphi_{ni}, [\ ])$ we get the method of least squares, while $L_i[\ ] = (\varphi_{ni}, [\ ])$ leads to Galerkin's method. Several other variants are frequently encountered. For example, if the underlying spaces involve functions defined on a region $D \in R^m$ and we define the functionals by

$$L_i f = f(t_i),$$

where the $t_i$ are distinct points in $D$, we get the *collocation method*. This method is often suggested because of its simplicity. Here $a_{ij} = (L\varphi_{nj})(t_i)$, so that the matrix elements are particularly simple to evaluate. There are, however, difficulties associated with this method. If we take $L = I$ and $\varphi_{ni} = t^{i-1}$, then the problem reduces to the approximation of a function $y$ by polynomial interpolation, and as we saw in Section 2.2 this process does not always converge. Thus we cannot expect the collocation method to work in all cases.

If we write $\chi_i = L\varphi_{ni}$, then (4.32) can be interpreted simply as the approximation of $y$ by a linear combination of the $\chi_i$ through general interpolation. The solubility of (4.32) can then be investigated using the results of Section 2.1. The question of convergence of the approximate solution is unfortunately a much more difficult matter. As indicated by the negative result on the collocation method we cannot treat (4.32) in full generality; if any positive results are to be obtained some restrictions have to be placed on the method. It is possible to get some results, but the analysis is quite complicated and we cannot pursue this further. The interested reader is referred to (Kantorovich and Akilov, 1964, Chapter 14), (Prenter, 1975), and a paper by Phillips (Phillips, 1972).

## Exercises 4.2

1. If the conditions of Theorem 4.9 are satisfied and if for each fixed $n$ the set $\{\varphi_{ni}\}$ is linearly independent, show that the minimization problem (4.24) has a unique solution.

2. Compute an improved approximation to the equation in Example 4.3 by including the term $\sin 3\pi t$ in the expansion. Compare the result with the known answer.

3. Consider the three-point boundary value problem

$$x''' - x = 0,$$
$$x(0) = x(1) = 0,$$
$$x(1/2) = 1.$$

   (a) Transform the problem so that the boundary conditions are homogeneous.

   (b) Find a simple polynomial satisfying the homogeneous boundary conditions and use the least squares method to get an approximate solution using a one term expansion.

   (c) Compare the approximate solution to the exact solution of the original problem.

4. Show that for Example 4.3 there exists a $\beta$ such that $\|Lx_n(\beta) - 1\|_2$ is arbitrarily small. Is it possible to find a $\beta$ such that $\|Lx_n(\beta) - 1\|_\infty$ is arbitrarily small?

5. Show how Theorems 4.9 and 4.10 can be modified (under appropriate assumptions) to eliminate the requirement that $L$ be bounded.

6. Use Galerkin's method to find an approximate solution of

$$x'' + tx = 1,$$
$$x(0) = x(1) = 0,$$

using $t(1-t)$ and $t^2(1-t)$ as expansion functions.

7. Prove that every positive-definite linear operator has a bounded inverse.

## 4.3  STABILITY AND CONVERGENCE

By now the astute reader may have noticed some potential difficulties with our approach. First of all, because of the assumptions made, the class of methods discussed in the previous section does not include a number of standard numerical algorithms. In particular, we assumed that the approximate solution lies in a subspace of the original space; such a restriction excludes methods like the finite-difference schemes for differential equations in which the approximate solution lies in $R^n$. This difficulty is easily overcome by a reformulation, as we see below.

A more serious objection concerns the actual solution of the approximating equations. Consider, for example, the least squares method. We have shown that as $n \to \infty$ the solution of (4.25) converges to the true solution provided the linear system can be solved exactly. But because of the inevitable presence of computational errors, one can only get an approximate solution of (4.25), which raises the question of the relation of the computed $x_n$ to the true solution. This is not a trivial concern; take, for instance, $L = I$, $\varphi_{ni} = t^{i-1}$ in $[0,1]$, and the usual inner product. Then the matrix $A$ in (4.25) is the $n \times n$ Hilbert matrix, which clearly limits the usefulness of the approach. What is needed is not only a theoretical proof of convergence but also some guarantee that the approximating equations are not too ill-conditioned. This leads us to the important concept of *stability*.

Consider again (4.1), always assuming that $L$ has a bounded inverse. In general, the approximate solution $x_n$, whatever its form, will be some element of a finite-dimensional space $X_n$ and (4.1) is replaced by

$$L_n x_n = y_n, \tag{4.35}$$

where $L_n$ is a linear operator $X_n \to X_n$. Since $X_n$ is not necessarily a subspace of $X$, we need to establish a connection between these space. A simple and elegant way of doing this is by introducing the so-called restriction and prolongation operators. For simplicity, let us assume that $Z$

is a Banach space, $L:X\to Y$, with $X$ and $Y$ subspaces of $Z$. The sequence of spaces $\{X_n\}$ will be assumed to be finite-dimensional with $\dim(X_n)=n$. A linear operator $r_n:Z\to X_n$ is then called a *restriction*, while a linear operator $p_n:X_n\to Z$ will be called a *prolongation*. We assume that $r_n$ and $p_n$ satisfy the following conditions:

$$\sup_n \|r_n\| \leq r < \infty, \qquad (4.36)$$

$$\sup_n \|p_n\| \leq p < \infty, \qquad (4.37)$$

$$r_n p_n = I_n, \qquad (4.38)$$

$$p_n r_n x \to x \qquad (4.39)$$

$$\|r_n x\| \to \|x\|. \qquad (4.40)$$

These conditions on $r_n$ and $p_n$ are quite natural and we assume, without further mention, that they are always satisfied.

**Example 4.5.** Take $Z=C[a,b]$ and $X_n=R^n$, both with maximum norm. Let $t_{n1},t_{n2},\ldots,t_{nn}$ be uniformly spaced points in $[a,b]$. Then we can define $r_n$ by

$$r_n x = \mathbf{x}_n = (x_{ni}) = \begin{bmatrix} x(t_{n1}) \\ x(t_{n2}) \\ \vdots \\ x(t_{nn}) \end{bmatrix}.$$

Linear interpolation provides a definition for $p_n$, that is, $p_n x_n$ is a piecewise linear function such that

$$(p_n x_n)(t_{ni}) = x_{ni}.$$

It is an easy exercise to show that all required conditions are satisfied (Exercise 4).

**DEFINITION 4.3.** A sequence of elements $x_n \in X_n$ is said to converge *discretely* to $x \in X$ if

$$\lim_{n\to\infty} \|x_n - r_n x\| = 0. \qquad (4.41)$$

The sequence $\{x_n\}$ is said to converge *globally* to $x$ if

$$\lim_{n\to\infty} \|x - p_n x_n\| = 0. \qquad (4.42)$$

With the assumption we have made the two types of convergence are equivalent.

**THEOREM 4.11.**   The sequence $\{x_n\}$ converges globally to $x$ if and only if it converges discretely to $x$.

PROOF.   Assume $p_n x_n \to x$. Then

$$x_n - r_n x = r_n p_n x_n - r_n x,$$
$$\|x_n - r_n x\| \leq \|r_n\| \|p_n x_n - x\|,$$

so that $\|x_n - r_n x\| \to 0$. Conversely, if we assume discrete convergence, then

$$x - p_n x_n = (x - p_n r_n x) + (p_n r_n x - p_n x_n),$$
$$\|x - p_n x_n\| \leq \|x - p_n r_n x\| + \|p_n\| \|r_n x - x_n\|.$$

Using (4.39) global convergence follows.    ∎

Since in the setting we have chosen the two types of convergence are equivalent we simply use the term convergence without qualification.

To investigate the approximations computed by (4.35) we consider the sequence of solutions $x_n$ as $n$ increases. With each $n$ we then associate a linear operator $L_n$, a restriction $r_n$, and a prolongation $p_n$. The sequence of triplets $\{L_n, r_n, p_n\}$ will be referred to as an *approximation* of $L$.

**DEFINITION 4.4.**   The approximation $\{L_n, r_n, p_n\}$ is said to be *consistent* with $L$ if for every $x \in X$

$$\lim_{n \to \infty} \|r_n L x - L_n r_n x\| = 0. \tag{4.43}$$

The expression

$$\delta_n(x) = r_n L x - L_n r_n x \tag{4.44}$$

is called the *consistency error* at $x$.

**DEFINITION 4.5.**   If for a given $x$, there exists a constant $c$, independent of $n$, such that

$$\|L_n r_n x - r_n L x\| \leq c n^{-k},$$

then the approximation $\{L_n, r_n, p_n\}$ is said to be *consistent of order $k$ with $L$ at $x$.

**DEFINITION 4.6.** A sequence of operators $\{L_n\}$ is said to be *stable* if the sequence $\{L_n^{-1}\}$ is uniformly bounded for sufficiently large $n$, that is,

$$\sup_{n \geq n_0} \|L_n^{-1}\| \leq k < \infty. \tag{4.45}$$

If the sequence $\{L_n\}$ is stable, then $\{L_n, p_n, r_n\}$ is a *stable approximation* of $L$.

It should be noted that stability is a property of the approximating scheme only. It is also worth remarking that the term "stability" is used because (4.45) implies that the solution of (4.35) is not overly sensitive to small perturbations; or at least it does not become increasingly more sensitive as $n \to \infty$. For a fuller exploration of this remark see Exercise 7.

The following two theorems, which connect the concepts of consistency, stability, and convergence, are perhaps the most important and general results in numerical analysis.

**THEOREM 4.12.** If $\{L_n, r_n, p_n\}$ is a stable approximation of $L$ then, for sufficiently large $n$

$$\|r_n x - x_n\| \leq k(\|\delta_n(x)\| + \|r_n y - y_n\|). \tag{4.46}$$

PROOF. Apply $r_n$ to (4.1), subtract from (4.35). Then

$$r_n L x - L_n x_n = r_n y - y_n,$$
$$L_n r_n x - L_n x_n = r_n y - y_n + L_n r_n x - r_n L x$$
$$r_n x - x_n = L_n^{-1}[r_n y - y_n - \delta_n(x)].$$

Applying the stability condition (4.45) completes the proof.  ∎

**THEOREM 4.13.** If $\{L_n, r_n, p_n\}$ is a stable and consistent approximation to $L$ and $y_n = r_n y$, then

(a)  $x_n$ converges to $x$.
(b)  The order of convergence is at least as large as the order of consistency at $x$.

PROOF. This follows immediately from (4.46).  ∎

In some cases it is desirable to replace stability by the somewhat less stringent notion of *weak stability*.

**DEFINITION 4.7.** An approximation $\{L_n, r_n, p_n\}$ is said to be *weakly stable* (of degree $r$) if for sufficiently large $n_0$,

$$\sup_{n \geq n_0} n^{-r} \|L_n^{-1}\| \leq k < \infty. \qquad (4.47)$$

Theorems 4.12 and 4.13 can then be modified in an obvious way to apply to weakly stable schemes. In particular, if a scheme is weakly stable of degree $r$ and consistent of order $k > r$, then the order of convergence of the approximate solution is at least $k - r$.

Theorems 4.12 and 4.13 appear in several forms in the literature. For initial value problems in ordinary and partial differential equations Theorem 4.13 is known as the *Lax Equivalence Theorem*. (In this particular instance it can be shown that stability is also necessary for convergence of a consistent scheme.)

In spite of the simplicity and generality of these results their application to particular problems is not always easy. Consistency is usually readily checked and most intuitively reasonable schemes are consistent. Stability presents a much more difficult problem. No universal scheme for verifying stability exists, but a variety of techniques, often depending on the nature of the particular problem, have been developed.

**Example 4.6.** Consider the two-point boundary value problem

$$Lu = -u'' + p(t)u' + q(t)u = y(t), \qquad (4.48)$$

$$u(a) = 0,$$

$$u(b) = 0,$$

where $|p(t)| \leq P$, $0 < Q_* \leq q(t) \leq Q^*$, $p, q \in C[a, b]$. It is known that under these assumptions the problem has a unique solution (Keller, 1968, Ch. 1).

A finite-difference method is constructed by introducing a mesh $a = t_{n1} < t_{n2} < \cdots < t_{nn} = b$, $t_{n,i+1} - t_{ni} = h$, and replacing the derivatives in (4.48) by centered differences. If we let $U_i$ denote the approximation to $u(t_{ni})$, the numerical solution is given by the system

$$\frac{2}{h^2}(a_i U_{i-1} + b_i U_i + c_i U_{i+1}) = y(t_{ni}), \qquad i = 2, \ldots, n - 1,$$

$$U_1 = 0, U_n = 0, \qquad (4.49)$$

where

$$a_i = -\frac{1}{2} - \frac{1}{4} hp(t_{ni}),$$

$$b_i = 1 + \frac{1}{2} h^2 q(t_{ni}),$$

$$c_i = -\frac{1}{2} + \frac{1}{4} hp(t_{ni}).$$

We define the restrictions $r_n$ by

$$r_n x = \begin{bmatrix} 0 \\ x(t_{n2}) \\ \vdots \\ x(t_{n,n-1}) \\ 0 \end{bmatrix}.$$

Then $L_n$ is a tridiagonal matrix of the form

$$L_n = \frac{2}{h^2} \begin{bmatrix} 1 & & \cdots & & & 0 \\ a_2 & b_2 & c_2 & & & \vdots \\ \vdots & a_3 & b_3 & c_3 & & \\ & & & \ddots & & \\ & & & a_{n-1} & b_{n-1} & c_{n-1} \\ 0 & & & & & 1 \end{bmatrix}.$$

Checking consistency is elementary, but the stability analysis requires more work. Consider the system

$$L_n z = \eta$$

where $z$ has components $z_i$ and $\eta$ components $\eta_i$. We claim that for $h \leq 2/P$,

$$\max|z_i| \leq \max(1, 1/Q_*)\max|\eta_i|. \tag{4.50}$$

To show this, assume that $\max|z_i|$ occurs at $2 \leq j \leq n-1$. Then

$$b_j z_j = -(a_j z_{j-1} + c_j z_{j+1}) + \frac{1}{2}h^2\eta_j,$$

$$\left(1 + \frac{1}{2}h^2 Q_*\right)|z_j| \leq |z_j| + \frac{1}{2}h^2|\eta_j|,$$

$$|z_j| \leq \frac{1}{Q_*}|\eta_j|.$$

Since $z_1 = \eta_1$ and $z_n = \eta_n$ (4.50) follows. Hence,

$$\|\mathbf{z}\|_\infty \leq \max(1, 1/Q_*)\|\boldsymbol{\eta}\|_\infty$$

and, by Theorem 4.7, $\|L_n^{-1}\| \leq \max(1, 1/Q_*)$, proving stability.

The method is consistent and stable, and therefore by Theorem 4.13 convergent. If the solution $u$ is sufficiently smooth we can show that the order of consistency, and hence the order of convergence, is two (Exercise 1).

The above proof of stability is an example of the use of the *maximum principle*, a technique often used in the theoretical and numerical analysis of ordinary and partial differential equations. Other commonly used approaches for proving stability utilize Fourier analysis, spectral analysis, or operator perturbation theory. A number of applications of these techniques to concrete problems can be found in Richtmyer and Morton (1967).

### Exercises 4.3

1. For Example 4.6 show that if $u \in C^{(4)}[a,b]$, then the order of consistency is two. Give explicit bounds for $\|L_n r_n u - r_n L u\|$.

2. Consider the elliptic partial differential equation

$$\frac{\partial^2 u}{\partial x^2} + \frac{\partial^2 u}{\partial y^2} - u = 0$$

in the unit square, with $u$ given on the boundary. A well-known approximation method consists of introducing a uniform grid on the square and replacing the partial derivatives by centered differences, using the standard 5-point scheme. Show that this method is stable. Under appropriate assumptions on $u$ show that the method is consistent and find the order of consistency.

3. Consider the Fredholm integral equation

$$x(t) + \int_0^1 K(t,s)x(s)\,ds = y(t), \qquad 0 \le t \le 1,$$

where $x(t)$ is an unknown function to be determined and $K(t,s)$ and $y(t)$ are given. Assume that $|K(t,s)| \le 1/2$ for $0 \le t \le 1$, $0 \le s \le 1$ and that $K \in C^{(\infty)}$ in both variables.

(a) Show that the equation has a unique solution.

(b) To solve this equation approximately we introduce mesh-points $0 = s_1 < \cdots < s_n = 1$, replace the integral by a numerical quadrature on these points, and satisfy the equation at the mesh-points. The approximate solution $z \in R^n$ then has components given by the solution of the system

$$z_i + \sum_{j=1}^n w_j K(s_i, s_j) z_j = y(s_i), \qquad i = 1, \ldots, n$$

where the $w_j$ are the weights of the chosen quadrature. If all $w_j \ge 0$, and the quadrature is exact for constants prove that the method is stable.

(c) Find the order of convergence if the composite Simpson's rule is used as quadrature.

4. Show that the restrictions and prolongations in Example 4.5 satisfy conditions (4.36)–(4.40).

5. Show how Theorems 4.12 and 4.13 can be extended for weakly stable schemes.

6. Investigate what changes are required in Example 4.6 if the boundary conditions are not homogeneous.

7. Mikhlin (1971, p. 57) defines stability as follows: Let $x_n$ be the solution of

$$L_n x_n = y_n$$

and $x_n + \Delta x_n$ be the solution of

$$(L_n + \Delta L_n)(x_n + \Delta x_n) = y_n + \Delta y_n$$

and assume $L_n$ is invertible. The method is then said to be stable if there exist constants $p, q, r$ independent of $n$, such that for $\|\Delta L_n\| \le r$ the perturbed equation is solvable for arbitrary $\Delta y_n$ and

$$\|\Delta x_n\| \le p\|\Delta L_n\| + q\|\Delta y_n\|.$$

Show that if a method is stable in the sense of Definition 4.6, then it is also stable in the Mikhlin sense.

8. Prove the following theorem: If for a given $L$ there exists a stable and consistent approximation $\{L_n, r_n, p_n\}$, then $L$ must have a bounded inverse.

## 4.4   ERROR ESTIMATES AND EXTRAPOLATION

Let us now briefly review what we have accomplished so far. The formulation (4.35) is general enough to include most standard algorithms for solving the linear inverse problem. The conceptually simple notions of consistency and stability give us a means for analyzing the convergence of a method, and stability implies that the approximate solution is not excessively sensitive to small computational errors. Thus a consistent and stable method is certainly a viable candidate for the solution of our problem. Furthermore, knowledge of the order of consistency and convergence give us a rule of thumb for judging efficiency; we tend to consider a higher order method as more efficient than a lower order one.

In practical situations such knowledge is reassuring, but not always sufficient. Once we have chosen a particular algorithm, selected a value for $n$, and solved (4.35) to obtain an approximate solution $x_n$, we are faced with the problem of bounding, or at least estimating, the error. We must now take a closer look at this problem to see what can be done.

In the proof of Theorem 4.12 we found that

$$\|r_n x - x_n\| \leq \|L_n^{-1}\|(\|L_n r_n x - r_n Lx\| + \|r_n y - y_n\|). \qquad (4.51)$$

$\|L_n^{-1}\|$ is certainly computable, but obtaining a bound on the rest of the expression on the right-hand side is difficult since $x$ is not known. In a few simple cases it is possible, through analytic manipulations, to bound this quantity, but even then (4.51) tends to give overly pessimistic results. In more complicated cases (4.51) is not useful at all for estimating the error.

Alternatively, we can use the prolongations $p_n$ and compute the trial solution $\hat{x} = p_n x_n$ with residual $r(\hat{x}) = L\hat{x} - y$. Then, from Theorem 4.8,

$$\|x - \hat{x}\| \leq \|L^{-1}\| \|r(\hat{x})\|. \qquad (4.52)$$

This requires knowledge of a bound on $\|L^{-1}\|$; often there are theoretical results that give us this information. If $r(\hat{x})$ is small, then (4.52) may give a realistic bound; unfortunately a good solution does not always result in a small residual. This difficulty can arise if $L$ is unbounded, for instance, for differential equations of high order. Here small differences between $x$ and

$\hat{x}$ may result in a large value for $L(x - \hat{x})$, so that (4.52) will give useless results.

While rigorous error bounds like the above are often computable for simple problems they tend to become unmanageable for even mildly complicated equations and one is forced to rely on more intuitive criteria. One commonly used technique is to solve (4.35) for some chosen $n$, then increase $n$ and solve the equation again. The difference in the two computed solutions is then taken as a measure of the error. Such a procedure cannot be defended with strictly rigorous arguments, since one can always find examples for which it leads to completely wrong estimates. Nevertheless, since the method does work in most practical cases it is of interest to analyze it so as to make it at least plausible. We will do this by investigating the error more closely. For simplicity we will assume for the rest of this chapter that $y_n = r_n y$. Our interest here is the detailed investigation of the *discretization error*

$$\epsilon_n = x_n - r_n x. \tag{4.53}$$

**DEFINITION 4.8.** The discretization error is said to have an *asymptotic expansion of order $p$* if there exists an $e \in X$ such that

$$\lim_{n \to \infty} \|n^p \epsilon_n - r_n e\| = 0. \tag{4.54}$$

It is of course true that if the method is consistent with order $p$ and stable, then $\|\epsilon_n\| = O(n^{-p})$, but (4.54) says more than that. If (4.54) holds, then the dominant part of the discretization error is smooth in some sense and its behavior with increasing $n$ is predictable. The existence of an asymptotic expansion for the discretization error depends on the existence of a similar expansion for the consistency error as we now show.

**DEFINITION 4.9.** The consistency error $\delta_n(x)$ is said to have an asymptotic expansion of order $p$ if there exists a linear operator $M : X \to Y$ such that

$$\lim_{n \to \infty} \|n^p \delta_n(x) - r_n M x\| = 0. \tag{4.55}$$

**THEOREM 4.14.** If the conditions of Theorem 4.13 are satisfied and if the consistency error $\delta_n(x)$ has an asymptotic expansion of order $p$ as defined in (4.55), then the discretization error has an asymptotic expansion of order $p$, with $e$ in (4.54) given as the solution of

$$Le = Mx. \tag{4.56}$$

PROOF.   We have from (4.55)

$$L_n \epsilon_n = \delta_n(x)$$
$$= n^{-P} r_n Mx + n^{-P} \xi_n,$$

with $\lim \|\xi_n\| = 0$. Introducing the scaled error $\eta_n = n^P \epsilon_n$ we have

$$L_n \eta_n = r_n Mx + \xi_n.$$

Consider now the equation

$$Le = Mx$$

and its approximate solution $e_n$ given by

$$L_n e_n = r_n Mx.$$

Because the method is stable and consistent, it follows from Theorem 4.13 that

$$\lim_{n \to \infty} \|e_n - r_n e\| = 0.$$

Now

$$L_n(\eta_n - e_n) = \xi_n$$

and from the stability of $\{L_n\}$

$$\|\eta_n - e_n\| \le k \|\xi_n\|.$$

Finally,

$$\|n^P \epsilon_n - r_n e\| = \|\eta_n - r_n e\|$$
$$\le \|\eta_n - e_n\| + \|e_n - r_n e\|,$$

completing the proof.    ∎

This theorem tells us that the discretization error behaves in a very predictable way; asymptotically it approaches $n^{-P} r_n e$, where $e$ satisfies an equation similar to the original equation, but with a different right-hand side. For sufficiently large $n$ this dominant part $n^{-P} r_n e$ will always be large compared to the neglected part, but the question of how large an $n$ this requires remains unanswered. The best one can do in practice is to assume that for the actual $n$ used the higher order terms in the error are negligible; in most actual problems one observes that this stage is reached even for moderate values of $n$. But because we are forced to make such unverifiable assumptions we must keep in mind that the techniques based on this theory must be used with caution.

A direct attempt to determine the dominant part of the discretization error by solving (4.56) is not possible, since the right-hand side involves the unknown solution $x$. Nevertheless, the information contained in Theorem 4.14 can be used in various ways, as we now show.

Let us assume that the conditions of Theorem 4.14 are satisfied. If we solve (4.35) with some value for $n$, we obtain an approximate solution $x_n \in X_n$ with error $\epsilon_n$. If we then take a different value for the discretization parameter, say $m = \alpha n$, $\alpha > 1$, we obtain a solution $x_m \in X_m$ with error $\epsilon_m$. In comparing the two solutions we must remember that $X_n$ and $X_m$ are not necessarily the same. To overcome this difficulty we introduce the linear operators $r_{mn} : X_m \to X_n$ and require that these operators be uniformly bounded and satisfy

$$r_{mn} r_m = r_n. \tag{4.57}$$

The meaning of the last requirement may not be obvious to the reader at this point, but reference to the concrete case discussed in Example 4.7 will be helpful. In the case where $X_n$ and $X_m$ are subspaces of $X$ we can take $r_n = I$, so that (4.57) is a trivial requirement.

From Theorem 4.14 it follows that

$$r_n x = x_n - \epsilon_n = x_n - n^{-p} r_n e + n^{-p} o(1), \tag{4.58}$$

$$r_m x = x_m - \epsilon_m = x_m - m^{-p} r_m e + m^{-p} o(1), \tag{4.59}$$

Applying $r_{mn}$ to the last equation, subtracting from (4.58) and using (4.57) we get

$$x_n - r_{mn} x_m = (n^{-p} - m^{-p}) r_n e + n^{-p} o(1). \tag{4.60}$$

The left-hand side is the difference (in $X_n$) between the two computed solutions; the first term on the right-hand side is of the same order of magnitude as the dominant part of the error. This provides some justification for taking the difference between the two computed solutions as an approximation to the discretization error. We can do even better: multiply (4.59) by $\alpha^p$, apply $r_{mn}$, and subtract from (4.58). Then

$$(1 - \alpha^p) r_n x = x_n - \alpha^p r_{mn} x_m + n^{-p} o(1).$$

Thus the extrapolated solution

$$x_n^e = (x_n - \alpha^p r_{mn} x_m)/(1 - \alpha^p) \tag{4.61}$$

differs from $r_n x$ by a term of order $n^{-p} o(1)$, so that this solution can be expected to be more accurate, provided of course, that the "asymptotic" stage has been reached, that is, that $n$ is sufficiently large. This very useful technique is called *Richardson's Extrapolation*.

**Example 4.7.**   Consider the two-point boundary value problem

$$u'' - u = 1,$$
$$u(0) = u(1) = 0.$$

Using the method and notation introduced in Example 4.6, the approximate solution is given by the system

$$\frac{U_{i-1} - 2U_i + U_{i+1}}{h^2} - U_i = 1, \qquad i = 2, \dots, n-1,$$

with $U = U_n = 0$. Since this is just a special case of the equation in Example 4.6 the scheme is stable. It follows then immediately that the consistency error is a vector $\delta$ with components

$$\delta_i = -\frac{u(t_{n,i-1}) - 2u(t_{ni}) + u(t_{n,i+1})}{h^2} + u''(t_{ni}).$$

If $u \in C^{(6)}[0,1]$ (which is the case here), then a straightforward Taylor expansion shows that

$$\delta_i = -\frac{1}{12} h^2 u^{(4)}(t_{ni}) + O(h^4).$$

The discretization error thus has an asymptotic expansion of the form (4.55) with $p = 2$ and

$$M[\ \ ] = -\frac{1}{12} \frac{d^4}{dt^4}[\ \ ]$$

so that we are justified in applying the extrapolation.

If we use just four points, relabelled for convenience $t_1, t_3, t_5, t_7$, and compute the solution $U$ numerically we obtain

$$U_3 = U_5 = -0.100000.$$

We now introduce the additional points $t_2, t_4, t_6$, midway between the previous points, and compute the new solution $\overline{U}$. This yields

$$\overline{U}_2 = \overline{U}_6 = -0.063322, \quad \overline{U}_3 = \overline{U}_5 = -0.100625, \quad \overline{U}_4 = -0.112945.$$

If we wish to extrapolate we can do so only at $t_3$ and $t_5$, because these are the only two points where we have two approximate solutions. Equation (4.57) is a formal statement of this requirement that the points of the finer mesh have to have as a subset the points of the coarser mesh. In this particular case the operators in (4.57) are defined by

$$
r_n x = \begin{bmatrix} x(t_1) \\ x(t_3) \\ x(t_5) \\ x(t_7) \end{bmatrix},
$$

$$
r_m x = \begin{bmatrix} x(t_1) \\ x(t_2) \\ \vdots \\ x(t_7) \end{bmatrix},
$$

$$
r_{mn} \begin{bmatrix} x(t_1) \\ x(t_2) \\ \vdots \\ x(t_7) \end{bmatrix} = \begin{bmatrix} x(t_1) \\ x(t_3) \\ x(t_5) \\ x(t_7) \end{bmatrix}.
$$

Extrapolation via (4.61) with $p=2$, $\alpha=2$ yields

$$U_3^e = U_5^e = -0.100833$$

in good agreement with the correct solution

$$u(t_3) = u(t_5) = -0.100836.$$

This example, while rather simple, demonstrates some of the properties and desirable features of Richardson's extrapolation:

1. The method is simple in application; one need only to solve the approximation equations twice with different (and appropriately

chosen) values of the discretization parameter. The actual extrapolation is trivial. In our example the method was found to be effective even for very low values of $n$. This is partly due to the fact that the problem is quite simple and its solution well-behaved, but often a similar behavior is observed in more complicated problems.

2.  The error can be estimated as part of the computation. From (4.60) we find that

$$|u(t_3) - \bar{U}_3| \lesssim 2 \times 10^{-4},$$

a good estimate for the error in $U_3$. The extrapolated value is considerably more accurate, but to be on the safe side we can use

$$|u(t_3) - U_3^e| \lesssim 2 \times 10^{-4}.$$

Such an additional "safety factor" often compensates for the omission of the higher order terms.

3.  Extrapolation often gives highly accurate results with little computational effort. In our example the error in the extrapolated solution can actually be shown to be $O(h^4)$. To prove this requires a simple extension of Theorem 4.14; this essentially involves a closer estimate of the neglected term $n^{-p}o(1)$ and is outlined in Exercise 1. One could of course devise higher order methods directly by replacing the derivatives by higher order differences, but this produces a number of difficulties. The stability analysis becomes more complicated, the resulting matrix is no longer tridiagonal, and centered differences cannot be used near the ends of the interval. Extrapolation allows us to construct higher order methods with a minimum of theoretical and practical difficulties.

The price one pays for all of these advantages is that in practical situations it is usually quite difficult to justify rigorously the use of extrapolation.

To investigate another possible way of estimating the error and to improve the solution, let us reconsider (4.56). As we have already remarked, this poses a problem since $x$ is unknown. Since we have, however, an approximation $x_n$ it is natural to ask whether it is possible to use this information to estimate $Mx$ and to use this estimate to compute $e$ approximately.

**THEOREM 4.15.**   If the conditions of Theorem 4.14 are satisfied, and if there exists an operator $U: X_n \to X_n$ such that

$$\lim_{n \to \infty} \|r_n Mx - Ux_n\| = 0, \tag{4.62}$$

then the corrected solution

$$x_n^c = x_n - n^{-p}L_n^{-1}Ux_n \tag{4.63}$$

satisfies

$$\lim_{n \to \infty} n^p \|x_n^c - r_n x\| = 0. \tag{4.64}$$

PROOF. Using the notation employed in Theorem 4.14 we can write

$$
\begin{aligned}
n^p(x_n^c - r_n x) &= n^p x_n - L_n^{-1}Ux_n - n^p r_n x \\
&= n^p x_n - L_n^{-1}r_n Mx - n^p r_n x - L_n^{-1}(Ux_n - r_n Mx) \\
&= n^p x_n - e_n - n^p r_n x - L_n^{-1}(Ux_n - r_n Mx) \\
&= (n^p \epsilon_n - e_n) - L_n^{-1}(Ux_n - r_n Mx).
\end{aligned}
$$

Applying the results obtained in Theorem 4.14 then proves (4.64). ■

The difficulty in applying this result lies primarily in verifying assumption (4.62).

**Example 4.8.** Consider again the simple equation in Example 4.7. We know that $r_n Mx$ has components

$$-\frac{1}{12}\frac{d^4}{dt^4}u(t_{ni}).$$

To approximate this we replace the fourth derivative by a five-point centered difference and $u$ by $U$. This suggests that

$$\frac{d^4u(t_{ni})}{dt^4} \simeq \frac{U_{i-2} - 4U_{i-1} + 6U_i - 4U_{i+1} + U_{i+2}}{h^4}. \tag{4.65}$$

Is such an approximation justified? We know that

$$U_i = u(t_{ni}) + h^2 e(t_{ni}) + h^2 o(1)$$

but this is not enough to show that (4.62) holds. We need to prove that the expansion can be carried out to more terms, in particular we want to show that

$$U_i = u(t_{ni}) + h^2 e_1(t_{ni}) + h^4 e_2(t_{ni}) + \text{higher order terms.}$$

For this we need to extend Theorem 4.14 to demonstrate the existence of a higher order expansion. With proper assumptions this can be done, but we shall omit the details. We will just give numerical results to show the effectiveness of the method for our example.

With mesh-points $t_0, t_1, t_2, t_3, t_4$ the initially computed solution is

$$U_1 = U_3 = 0.084922, \quad U_2 = -0.112652.$$

Substituting these results in (4.65) and carrying out the correction, the new results are

$$U_1^c = U_3^c = -0.085314, \quad U_2^c = -0.1133173$$

which compares well with the exact solution

$$u(t_1) = u(t_3) = -0.085323, \quad U(t_2) = -0.1133181.$$

It can be shown that the approximation solutions computed in this way converge to the true solution with order four.

The general procedure defined by (4.63) is called the *method of deferred corrections*. It has the advantage over Richardson's extrapolation in that we need only solve the same equation twice, with different right-hand sides, whereas for extrapolation the approximation equation has to be solved with two different values of $n$. This advantage is offset by the fact that more manipulations are required to get $Ux_n$ and that (4.62) is complicated to verify. Generally speaking, the method of deferred corrections tends to be more efficient computationally but harder to apply than Richardson's extrapolation.

### Exercises 4.4

1. Modify the proof of Theorem 4.14 to show that if

$$\delta_n(x) = n^{-p} r_n M x + O(n^{-\bar{p}}), \quad \bar{p} > p,$$

then the discretization error has an expansion of the form

$$\| \epsilon_n - n^{-p} r_n e \| = O(n^{-p^*}),$$

with $p^* = \min(2p, p + \bar{p})$, provided $Mx$ satisfies certain additional assumptions (state what these should be).

2. Assume that the consistency error has an expansion of the form

$$\delta_n(x) = n^{-p_1} r_n M_1 x + n^{-p_2} r_n M_2 x + O(n^{-p_3})$$

with $p_1 < p_2 < p_3$. Show that the discretization error has the form

$$\epsilon_n = n^{-p_1^*} r_n e_1 + n^{-p_2^*} r_n e_2 + O(n^{-p_3^*}).$$

Find expressions for $p_1^*, p_2^*, p_3^*$ and the equations satisfied by $e_1$ and $e_2$.

3. Modify Theorem 4.15 to show that if

$$\|r_n Mx - Ux_n\| = O(n^q), \qquad q > 1$$

and if $\delta_n(x)$ is as in Exercise 1, then

$$\|x_n^c - r_n x\| = O(n^{-Q}),$$

with $Q = \min(p + \bar{p}, p + q)$.

4. Consider the Laplace equation

$$\frac{\partial^2 u}{\partial x^2} + \frac{\partial^2 u}{\partial y^2} = 0,$$

in the unit square with boundary conditions

$$u(x,0) = 0, \quad u(0,y) = \sin y,$$
$$u(1,y) = e \cdot \sin y, \quad u(x,1) = e^x \sin 1.$$

Solve this problem using a finite-difference method based on the five-point difference scheme. Using the minimum number of mesh-points necessary, compute the approximate solution, then apply the method of deferred corrections. Compare your results with the known answer $u(x,y) = e^x \sin y$.

## 4.5 APPROXIMATE COMPUTATION OF EIGENVALUES

So far in this chapter we have considered problems with a unique solution, so that the homogeneous case $Lx = 0$ has a trivial solution only. We now briefly look at the case when this is not so, in particular we consider the standard eigenproblem

$$(L - \lambda I)x = 0. \tag{4.66}$$

A $\lambda$ for which this equation has a nontrivial solution is an *eigenvalue* of $L$; the corresponding $x$ is the *eigenvector* associated with $\lambda$. The set of all eigenvalues of $L$ will be called the *spectrum* of $L$ and denoted by $\sigma(L)$. Actually, in the usual terminology of functional analysis, the set of eigenvalues is called the point spectrum, the complete spectrum is taken to include some other values of $\lambda$. We are interested here only in eigenvalues and we use the term spectrum accordingly.

The theory of the approximate solution of the eigenproblem is still undergoing development and seems not yet to have reached the stage where an elementary treatment is possible for the general case. For this reason we will discuss only some isolated ideas; these are representative of the types of results that have been obtained, but do not form a complete theory by any means. For more details the reader will need to consult the literature.

The discussion of the eigenproblem is simplified considerably if we assume that the underlying space is a Hilbert space and that $L$ is linear and symmetric.

**DEFINITION 4.10.**   A linear operator $L$ from a Hilbert space $X$ into $X$ is said to be symmetric if

$$(Lx,y) = (x,Ly) \tag{4.67}$$

for all $x,y \in \mathcal{D}(L)$.

The following theorem is standard.

**THEOREM 4.16.**   If $L$ is a bounded symmetric operator on a Hilbert space $X$ with $\mathcal{D}(L) = X$, then

$$\|L\| = \sup_{\lambda \in \sigma(L)} |\lambda|. \tag{4.68}$$

PROOF.   See Taylor (1964, Ch. 6).   ∎

For the rest of this section we will make the following assumptions:

1.  $X$ is a Hilbert space, $L$ is a linear symmetric operator with $\mathcal{D}(L)$ and $\mathcal{R}(L)$ in $X$.

2.  $\sigma(L)$ contains at most a countably infinite set of real eigenvalues, having no finite accumulation point except possibly zero.

3.  To each $\lambda \in \sigma(L)$ correspond one or more eigenvectors. The value $k = \dim(\mathcal{N}(L - \lambda I))$ is finite and is called the multiplicity of the eigenvalue $\lambda$.

4.  If $\mu \notin \sigma(L)$ and $\mu \neq 0$, then $(L - \mu I)$ has a bounded inverse on $X$.

These assumptions are of course not all independent; to see their relationship the reader can consult the section dealing with spectral analysis in any book on functional analysis.

**THEOREM 4.17.**   For given $\mu$ and $z \in \mathcal{D}(L)$, let

$$r = Lz - \mu z.$$

Then there is a $\lambda \in \sigma(L)$ such that

$$|\mu - \lambda| \leq \|r\|/\|z\|. \tag{4.69}$$

PROOF.   If $\mu \in \sigma(L)$, then there is nothing to prove. Otherwise $L - \mu I$ has a bounded inverse so that

$$z = (L - \mu I)^{-1} r$$

$$\|z\| \leq \|(L - \mu I)^{-1}\| \, \|r\|.$$

From Theorem 4.16 and Exercises 1 and 2

$$\|(L - \mu I)^{-1}\| = \sup_{\lambda \in \sigma(L)} |(\lambda - \mu)^{-1}|$$

$$= \frac{1}{\displaystyle\inf_{\lambda \in \sigma(L)} |\lambda - \mu|}.$$

Hence

$$\inf_{\lambda \in \sigma(L)} |\lambda - \mu| \leq \|r\|/\|z\|$$

giving (4.69).   ■

This theorem shows that a small residual implies that $\mu$ is close to an eigenvalue, an indication that the assumptions we have chosen make the problem well-posed in some sense. This can also be seen by considering the effect of a small perturbation on the spectrum.

**THEOREM 4.18.**   Let $\lambda \in \sigma(L)$, with corresponding normalized eigenvector $x$. If $\Delta L$ is a bounded symmetric operator with $\mathcal{D}(\Delta L) = \mathcal{D}(L)$, then $\sigma(L + \Delta L)$ contains at least one point $\bar{\lambda}$ in the interval $[\lambda - \|\Delta L\|, \lambda + \|\Delta L\|]$.

PROOF.   Let $r = (L + \Delta L)x - \lambda x$. Then, since $\|x\| = 1$, $\|r\| \leq \|\Delta L\|$ and the result follows from Theorem 4.17.   ■

The last result is of course incomplete since it does not consider whether $L$ and $L + \Delta L$ have the same number of eigenvalues (with the same multiplicities) in some fixed interval on the real line. We do not pursue this more difficult question of establishing a one-to-one correspondence between the two spectra.

To solve the eigenproblem numerically we again have to resort to discretization, leading, in general, to an algebraic eigenvalue problem, that can then be solved by any of the known algorithms. The discretization is accomplished using the same principles that were employed in the solution of the inverse problem. Thus (4.66) is replaced by

$$(L_n - \lambda_n I_n) x_n = 0 \qquad (4.70)$$

where $x_n$ is an element of finite-dimensional space $x_n$ and $L_n : X_n \to X_n$ is symmetric*. Again, to connect the spaces $X$ and $X_n$ we use restrictions and prolongations, and proceed in a manner similar to Section 4.4. We use all notation, definitions, and assumptions as given in that section.

**THEOREM 4.19.** Let $\lambda$ be an eigenvalue of $L$ with corresponding eigenvector $x$. Then $\sigma(L_n)$ contains at least one point $\lambda_n$ such that

$$|\lambda - \lambda_n| \leq \|\delta_n(x)\| / \|r_n x\|, \qquad (4.71)$$

where $\delta_n(x)$ is the consistency error as defined in (4.44).

PROOF. From (4.66)

$$r_n L x - \lambda r_n x = 0,$$

so that

$$L_n r_n x - \lambda r_n x = -\delta_n(x).$$

Since $L_n$ is symmetric we can apply Theorem 4.17 and, with $\mu = \lambda$, $\lambda = \lambda_n$, (4.71) follows. ∎

Note that $\delta_n(x)$ is simply the consistency error for the eigenvector $x$. If the approximation is consistent with $L$ and since by (4.40)

$$\|r_n x\| \not\to 0,$$

we see that the sequence of approximate eigenvalues $\{\lambda_n\}$ converges to $\lambda$.

---

*In some cases discretization of a symmetric operator does not directly lead to a symmetric matrix, but in such cases it is usually possible to find a simple similarity transformation which symmetrizes the matrix. Since our conclusions are not affected by this slight complication, we will ignore this point.

Of course, we must keep in mind what the convergence means: if the original problem has an eigenvalue $\lambda$ then, for sufficiently large $n$, the approximate problem will have at least one eigenvalue arbitrarily close to $\lambda$. We may also note that the difference between $\lambda$ and $\lambda_n$ is essentially proportional to the consistency error for the eigenvector $x$, so that the order of convergence is the same as the order of consistency. Equation (4.71) also explains the empirical observation that eigenvalues associated with smooth eigenvectors are usually much better approximated than those associated with highly oscillatory eigenvectors, since in the former case the consistency error tends to be much smaller.

**Example 4.9.** The Galerkin method introduced in Section 4.4 can be adapted for the solution of the eigenvalue problem. Let $\varphi_1, \varphi_2, \ldots$ be a set of elements closed in $X$. For simplicity we assume also that they are orthonormal. Let us then find an

$$x_n = \sum_{j=1}^{n} \alpha_j \varphi_j$$

such that, for certain values of $\lambda$, the residual

$$r(x_n) = Lx_n - \lambda x_n$$

is orthogonal to $\varphi_1, \varphi_2, \ldots, \varphi_n$. These conditions will determine the $\alpha_j$ and a set of possible $\lambda$.

For the orthogonality relation to be satisfied we need

$$(r, \varphi_i) = \left( \sum_{j=1}^{n} \alpha_j L\varphi_j - \lambda \varphi_j, \varphi_i \right) = 0, \qquad i = 1, 2, \ldots, n,$$

or, because of the orthonormality of the $\varphi_i$,

$$\sum_{j=1}^{n} \alpha_j \big( (L\varphi_j, \varphi_i) - \lambda \delta_{ij} \big) = 0. \tag{4.72}$$

This is of course a standard matrix eigenvalue problem for the symmetric matrix $A$ with elements

$$a_{ij} = (L\varphi_j, \varphi_i).$$

To analyze the method we can consider it a special case of (4.70) with

$X_n = R^n$ and restrictions defined by

$$r_n x = \begin{pmatrix} (x, \varphi_1) \\ (x, \varphi_2) \\ \vdots \\ (x, \varphi_n) \end{pmatrix}.$$

If we write

$$x = \sum_{j=1}^{\infty} (x, \varphi_j) \varphi_j,$$

then $r_n L x$ and $L_n r_n x$ are vectors with components

$$(r_n L x)_i = \sum_{j=1}^{\infty} (x, \varphi_j)(L\varphi_j, \varphi_i),$$

$$(L_n r_n x)_i = \sum_{j=1}^{n} (x, \varphi_j)(L\varphi_j, \varphi_i).$$

The method is then consistent if

$$\lim_{n \to \infty} \sup_{1 \le i \le n} \sum_{j=n}^{\infty} (x, \varphi_j)(L\varphi_j, \varphi_i) = 0.$$

For this to be true, it is sufficient, but not necessary, that $L$ be bounded.

**Example 4.10.**   Consider the Sturm–Liouville problem

$$((1+t)u')' + \lambda u/(1+t) = 0,$$
$$u(0) = u(1) = 0,$$

whose exact eigenvalues are $\lambda_k = (k\pi/\log_e 2)^2$.

A finite difference method is constructed by introducing the uniform mesh of width $h$ on $[0,1]$ and replacing the derivatives by centered differences as in Example 4.6. The resulting matrix eigenvalue problem is then easily solved for the approximate eigenvalues. Numerical results for the smallest eigenvalue $\lambda = 20.54$ are given below for several values of $n$.

|  | $n=4$ | $n=8$ | $n=24$ |
|---|---|---|---|
| Computed $\lambda_n$ | 19.60 | 20.30 | 20.51 |
| $\lvert$Error$\rvert$ | 0.94 | 0.24 | 0.03 |

The method has an order of consistency of two, and the computed values bear out the predicted second-order convergence.

While Theorem 4.19 tells us that, for a consistent approximation, every eigenvalue in $\sigma(L)$ is approximated by some $\lambda_n$, it does not rule out the possibility that there may appear "spurious" eigenvalues, that is, that there may exist a sequence $\{\lambda_n\}$ converging to $\lambda \notin \sigma(L)$. One might be tempted to conjecture at this point that the assumptions made, namely $L$ and $L_n$ symmetric and $L_n$ consistent with $L$, might be sufficient to rule out such extraneous eigenvalues. However, this conjecture is false as we can see from a simple example.

**Example 4.11.**   Let $X$ be $l_2$ with the usual inner product

$$(x,y) = \sum_{i=1}^{\infty} x_i y_i.$$

Let $e_i$ denote the orthonormal unit vectors, $e_i = (0,0,\ldots,1,\ldots,0)$. Define the operator $L$ by

$$Lx = x_1 e_1$$

then $L$ is clearly bounded and symmetric. The only nonzero eigenvalue of $L$ is 1 with eigenvector $e_1$. Now take $X_n = l_2$ and $r_n = I$, and define $L_n$ as

$$L_n x = x_1 e_1 - x_n e_n.$$

Each $L_n$ is symmetric and $\sigma(L_n) = \{-1,0,1\}$. But $\{L_n\}$ is consistent with $L$ since for each $x \in l_2$

$$\|L_n x - Lx\| = |x_n| \to 0 \qquad \text{as } n \to \infty.$$

The eigenvalue $-1$ has been introduced spuriously by the discretization.

In order to rule out such spurious sequences we must then impose some additional conditions on $\{L_n\}$. We do this by defining the concept of stability for the eigenvalue problem; this notion is closely related to the concept already introduced in Section 4.3.

**DEFINITION 4.11.**   An approximation method $\{L_n, r_n, p_n\}$ is *stable for the solution of the eigenvalue problem* if for every $\mu \notin \sigma(L)$, the approximation $\{L_n - \mu I_n, r_n, p_n\}$ is stable approximation in the sense of Definition 4.6.

To rephrase this, a method which is stable for the solution of the inverse problem whenever the operator has a bounded inverse is also stable for the solution of the eigenvalue problem. If the approximation scheme is stable, then there are no spurious eigenvalues.

**THEOREM 4.20.** Let $\{L_n, r_n, p_n\}$ be consistent with $L$ and stable for the solution of the eigenvalue problem. If $\lambda \notin \sigma(L)$, then there exists an $\epsilon > 0$ such that, for sufficiently large $n$, $\sigma(L_n) \cap (\lambda - \epsilon, \lambda + \epsilon)$ is empty.

PROOF.    Assume that there exists a $\lambda_n \in \sigma(L_n)$ violating the theorem. Then

$$L_n x_n - \lambda_n x_n = 0$$

with $\|x_n\| = 1$. From this, we have

$$(L_n x_n - \lambda I_n) x_n = (\lambda_n - \lambda) x_n,$$

and by assumptions $\lambda \notin \sigma(L)$ and stability

$$\|x_n\| \le \|(L_n - \lambda I_n)^{-1}\| \, |\lambda_n - \lambda| \, \|x_n\|.$$

Therefore we must have

$$1 \le K |\lambda_n - \lambda| \le K\epsilon.$$

By making $\epsilon$ sufficiently small we obtain a contradiction.    ∎

In conclusion we point out again that our analysis is by no means complete; we have simply ignored the more difficult aspects. Typical of these is the question of the proper representation of the multiplicities in the approximate solution. If $\lambda$ is an eigenvalue of multiplicity $k$, what can we say about $\lambda_n$? Also, we have not considered the approximate computation of eigenvectors or the case when $L$ is not symmetric. Some answers to these questions are known, but the treatment is complicated. The interested reader may consult a paper by Chatelin (1973) and the references cited therein.

Another interesting question is raised by Example 4.10 which shows that the computed eigenvalue behaves predictably with increasing $n$, suggesting the possible application of Richardson's extrapolation or the method of deferred corrections. Some isolated results on this have been obtained, but a general theory does not yet exist.

**Exercises 4.5**

1. If $L$ is symmetric and $\mu \notin \sigma(L)$, show that $(L - \mu I)^{-1}$ is symmetric.
2. Prove the following statements:
   (a) If $\lambda, \mu \in \sigma(L)$ with $\lambda \neq \mu$, then the corresponding eigenvectors are orthogonal.
   (b) If $\sigma(L) = \{\lambda\}$, then $\sigma(L - \mu I) = \{\lambda - \mu\}$.
   (c) If $\mu \neq \sigma(L)$, then $\sigma((L - \mu I)^{-1}) = \{1/(\lambda - \mu)\}$.
3. Let $L$ have eigenvalues $\lambda_1, \lambda_2, \ldots$ with corresponding eigenvectors $x_1, x_2, \ldots$. Assume that $\lambda_k$ is a simple eigenvalue and that $\min_{i \neq k} |\lambda_k - \lambda_i| = a$. Let $\bar{\lambda}$ be an eigenvalue of $L + \Delta L$ approximating $\lambda_k$ in the sense of Theorem 4.18 and $\bar{x}$ the corresponding eigenvector. Show that if $\|\Delta L\| < a$, then

$$|(x_i, \bar{x})| \leq \|\Delta L\|/(a - \|\Delta L\|), \qquad \text{for all } i \neq k.$$

4. If $L$ does not have an eigenvalue in $[\mu, \mu + a]$, and if $\|\Delta L\| < a$, show that $L + \Delta L$ does not have an eigenvalue in $[\mu + \|\Delta L\|, \mu + a - \|\Delta L\|]$.
5. Consider the integral equation eigenvalue problem

$$Lx = \int_0^1 e^{st} x(s)\, ds = \lambda x(t), \qquad 0 \leq t \leq 1.$$

   (a) Find a bound for the spectral radius $\rho(L) = \sup_{\lambda \in \sigma(L)} |\lambda|$.
   (b) Devise a crude numerical method for estimating the eigenvalues by replacing the integral by the trapezoidal rule. Compare the two eigenvalues so obtained with the largest two eigenvalues of the original problem which are known to be 1.353 and 0.106.
   (c) What is the form of the matrix if the composite Simpson's rule is used to replace the integral? Find the similarity transformation which symmetrizes this matrix.

# 5

# THE NONLINEAR INVERSE PROBLEM

Let us now consider the more difficult inverse problem

$$Rx = y, \tag{5.1}$$

where $R$ is a nonlinear operator. For convenience we will consider the two alternate forms

$$x = Tx \tag{5.2}$$

and

$$Px = 0. \tag{5.3}$$

Equation (5.1) can always be rewritten in this fashion, for instance, by setting $T[\ ] = (I + R)[\ ] - y$ and $P[\ ] = R[\ ] - y$. Actually, in each specific instance, there are usually many ways in which (5.1) can be so reformulated; some of these will be found to be more useful than others.

The analysis carried out in the previous chapter depends very heavily on the fact that the operators are linear, so an immediate generalization is not possible and we must develop a new approach for the nonlinear problem. Of course, any conclusions we can draw about the nonlinear problem will be true for the linear case as well.

## 5.1 THE METHOD OF SUCCESSIVE SUBSTITUTIONS

In developing general methods it is often appropriate to proceed by analogy: if we have a procedure for solving a simple example we may try to extend it to handle the general case. Now the simplest instance of (5.1) is the problem of solving a nonlinear equation for a function of one variable. Techniques for this, often called "root-finding" methods, are a standard topic in introductory numerical analysis. Several of these methods can be generalized for operator equations.

One of the simplest root-finding techniques is the method of *successive substitutions*: to find the roots of $f(t) = 0$ we rewrite the equation as

$$t = g(t) \tag{5.4}$$

and, starting with some $t_0$, generate the sequence

$$t_n = g(t_{n-1}). \tag{5.5}$$

This sequence, if it converges at all, converges to a solution of (5.4).

Formally, we can carry this procedure over to general spaces; thus to solve (5.2) we generate the sequence

$$x_n = Tx_{n-1} \tag{5.6}$$

with some appropriately chosen initial guess $x_0$. In certain circumstances this produces a sequence converging to a solution of (5.2). The situation for the linear case was discussed in Chapter 4.

**DEFINITION 5.1.** Let $T$ be an operator mapping a linear space $X$ into itself. Then $x \in X$ is called a *fixed point* of $T$ if

$$x = Tx. \tag{5.7}$$

**THEOREM 5.1.** Let $T: X \to X$, with $X$ a normed linear space, and assume that the sequence $\{x_n\}$ generated by (5.6) converges to some $x^* \in X$. If $T$ is continuous at $x^*$, then $x^*$ is a fixed point of $T$.

**PROOF.**

$$Tx^* - x^* = (Tx^* - Tx_{n-1}) + (Tx_{n-1} - x_n) + (x_n - x^*),$$
$$\|Tx^* - x^*\| \le \|Tx^* - Tx_{n-1}\| + \|x_n - x^*\|.$$

Since $x_n \to x^*$, the second term on the right-hand side goes to zero as $n \to \infty$. Also, since $T$ is continuous at $x^*$ and $x_n \to x^*$, the first term tends to zero in the limit. Hence

$$\|Tx^* - x^*\| = 0. \qquad \blacksquare$$

Of course (5.6) does not always define a convergent sequence and the method of successive substitutions is useful only when certain conditions are fulfilled.

**DEFINITION 5.2.**   Let $b(z,r)$ be a neighborhood of $z \in X$ defined by

$$b(z,r) = \{x | \|x - z\| \le r\} \tag{5.8}$$

that is, $b(z,r)$ is the closed ball with radius $r$ and center $z$ in the topology defined by the norm. Then an operator $T: X \rightarrow X$ is said to be a *contraction mapping* in $b(z,r)$ if there exists a constant $0 \le \theta < 1$ such that

$$\|Tx_1 - Tx_2\| \le \theta \|x_1 - x_2\| \tag{5.9}$$

for all $x_1, x_2 \in b(z,r)$. $\theta$ is the *contraction factor* of $T$ in $b(z,r)$.

**Example 5.1.**   A linear operator $L$ with $\|L\| < 1$ is a contraction mapping for the whole space. Similarly, any bounded operator with bound less than one is a contraction mapping. In general, a nonlinear operator may be a contraction mapping in some subset of the space, but not on the whole space.

**Example 5.2.**   Let $X = R^2$ with the maximum norm. Then $T$ defined by

$$T\begin{pmatrix} x_1 \\ x_2 \end{pmatrix} = \frac{1}{2}\begin{pmatrix} x_1 + x_2^2 \\ x_1^2 + x_2 \end{pmatrix}$$

is a contraction mapping in $b(0,r)$ for $r \le 1/2$, but not a contraction mapping for $r > 1/2$.

**THEOREM 5.2.   THE CONTRACTION MAPPING THEOREM.**   Let $T$ be an operator from a Banach space $X$ into itself. Assume that $T$ is a contraction mapping with contraction factor $\theta < 1$ in $b(x_0, r)$ with

$$r \ge r_0 = \frac{1}{1-\theta} \|x_0 - Tx_0\|. \tag{5.10}$$

Then

(a)   $T$ has a unique fixed point $x^* \in b(x_0, r_0)$.
(b)   The sequence $x_n = Tx_{n-1}$, $n = 1, 2, \ldots$, converges to $x^*$.
(c)   $\|x_n - x^*\| \le \theta^n r_0$. $\tag{5.11}$

PROOF.   First we show that the sequence $\{x_n\} \in b(x_0, r_0)$. Assume that for $x_1, x_2, \ldots, x_n$ we have

$$\|x_n - x_0\| \le (1 - \theta^n) r_0 < r_0. \tag{5.12}$$

Then

$$\|x_{n+1} - x_n\| = \|Tx_n - Tx_{n-1}\| \le \theta \|x_n - x_{n-1}\|$$
$$\le \theta \|Tx_{n-1} - Tx_{n-2}\|$$
$$\le \theta^2 \|x_{n-1} - x_{n-2}\|$$
$$\vdots$$
$$\le \theta^n \|x_1 - x_0\|.$$

Thus

$$\|x_{n+1} - x_n\| \le \theta^n (1 - \theta) r_0 \tag{5.13}$$

and

$$\|x_{n+1} - x_0\| \le \|x_{n+1} - x_n\| + \|x_n - x_0\|$$
$$\le \theta^n (1 - \theta) r_0 + (1 - \theta^n) r_0$$
$$= (1 - \theta^{n+1}) r_0 < r_0.$$

Since

$$\|x_1 - x_0\| = \|Tx_0 - x_0\| = (1 - \theta) r_0,$$

(5.12) holds for $n = 1$ and hence by induction for all $n$, proving that all $x_n \in b(x_0, r_0)$.

Next we show that the sequence $\{x_n\}$ converges to $x^* \in X$.

$$\|x_{m+p} - x_m\| \le \|x_{m+p} - x_{m+p-1}\| + \|x_{m+p-1} - x_{m+p-2}\| + \cdots$$
$$+ \|x_{m+1} - x_m\|,$$

so that, from (5.13),

$$\|x_{m+p} - x_m\| \le \theta^{m+p-1}(1 - \theta) r_0 + \theta^{m+p-2}(1 - \theta) r_0 + \cdots$$
$$+ \theta^m (1 - \theta) r_0$$
$$= \theta^m (1 - \theta)(1 + \theta + \theta^2 + \cdots + \theta^{p-1}) r_0$$
$$\le \theta^m (1 - \theta) \frac{1}{1 - \theta} r_0 = \theta^m r_0. \tag{5.14}$$

Therefore

$$\lim_{m\to\infty} \lim_{p\to\infty} \|x_{m+p} - x_m\| = 0,$$

and $\{x_n\}$ is a Cauchy sequence. Since $X$ is complete, the sequence $\{x_n\}$ converges to $x^* \in X$. The inequality (5.11) follows by letting $p\to\infty$. Also, since $T$ is a contraction mapping in $b(x_0, r_0)$ it is continuous in the interior; hence by Theorem 5.1 $x^*$ is a fixed point of $T$.

To show that $x^*$ is unique, assume that there is another fixed point $x^{**} \neq x^*$ in $b(x_0, r_0)$. Then

$$\|x^* - x^{**}\| = \|Tx^* - Tx^{**}\| < \|x^* - x^{**}\|,$$

which is a contradiction. Hence $x^*$ is the only fixed point in $b(x_0, r_0)$. ∎

**Example 5.3.**   Find the solution of the nonlinear system

$$x + \tfrac{1}{4}y^2 = \tfrac{1}{16}$$

$$\tfrac{1}{3}\sin x + y = \tfrac{1}{2}.$$

Even a simple system like this can be written in the form of (5.2) in many ways, but not all of these will lead to a convergent algorithm. To get a contraction mapping we want the right-hand side of (5.2) to be "small" in some sense. In this example it is fairly obvious what one should do and we write

$$x_{n+1} = \tfrac{1}{16} - \tfrac{1}{4}y_n^2$$

$$y_{n+1} = \tfrac{1}{2} - \tfrac{1}{3}\sin x_n.$$

With

$$z_0 = \begin{pmatrix} x_0 \\ y_0 \end{pmatrix} = \begin{pmatrix} 0 \\ 0 \end{pmatrix}$$

we obtain the iterates

$$z_1 = \begin{bmatrix} \tfrac{1}{16} \\ \tfrac{1}{2} \end{bmatrix}, \quad z_2 = \begin{pmatrix} 0 \\ 0.4792 \end{pmatrix}, \quad z_3 = \begin{pmatrix} 0.0051 \\ 0.4983 \end{pmatrix}.$$

Using the maximum norm

$$\|z_0 - Tz_0\| = \|z_0 - z_1\| = 0.5.$$

Let us try the value $r = 1$. For $z_i, z_j \in b(0, 1)$

$$\| Tz_i - Tz_j \| = \max\left(\tfrac{1}{4}|y_i^2 - y_j^2|, \tfrac{1}{3}|\sin x_i - \sin x_j|\right)$$
$$\leq \max\left(\tfrac{1}{2}|y_i - y_j|, \tfrac{1}{3}|x_i - x_j|\right)$$
$$\leq \tfrac{1}{2}\|z_i - z_j\|.$$

Therefore (5.10) is satisfied with $\theta = \tfrac{1}{2}$. The system thus has a unique solution in $-1 \leq x \leq 1$, $-1 \leq y \leq 1$, and the iteration converges to this solution (which is of course $x = 0$, $y = \tfrac{1}{2}$). We should note, however, that this does not tell us anything about the possibility of other solutions outside $b(0, 1)$.

**Example 5.4.** Consider the system

$$x - 2y^2 = -1$$
$$3x^2 - y = 2$$

which has a solution $(1, 1)$. The iterative method

$$x_{n+1} = 2y_n^2 - 1,$$
$$y_{n+1} = 3x_n^2 - 2,$$

is not a contraction mapping near $(1, 1)$ and the procedure does not converge even if the starting value is quite close to the solution. On the other hand, the rearrangement

$$x_{n+1} = \sqrt{(y_n + 2)/3}$$
$$y_{n+1} = \sqrt{(x_n + 1)/2}$$

is a contraction mapping near the solution and the iterates converge if the starting guess is sufficiently good.

If $T$ is a contraction mapping near the solution $x^*$, then for $x_0$ near $x^*$ the sequence (5.6) will converge to $x^*$. The set of all starting points $x_0$ for which we obtain convergence to $x^*$ is called the *region of accessibility* for $x^*$. For the method of successive substitutions to converge we therefore must write the equation in the form of (5.2) in such a way that $T$ is a contraction mapping near the solution and then find a point $x_0$ in the region of accessibility. This requirement, though simply stated, is not always readily fulfilled in complicated cases. It is often difficult to find the region of accessibility and the situation can be quite complicated even in simple problems.

**Example 5.5.** The system

$$x + 0.25y^2 = 1.25$$
$$0.25x^2 + y = 1.25$$

has two solutions, $(1,1)$ and $(-5,-5)$. The iterative procedure

$$x_{n+1} = 1.25 - 0.25y_n^2$$
$$y_{n+1} = 1.25 - 0.25x_n^2$$

is a contraction mapping near $(1,1)$, hence the region of accessibility for $(1,1)$ contains some neighborhood of this point. At $(-5,-5)$ the mapping is not a contraction and the sequence does not converge to $(-5,-5)$ for any starting value near this point. However, the starting values $(5,5)$, $(5,-5),(-5,5),(-5,-5)$ yield $(-5,-5)$ in one iteration. Hence the region of accessibility for $(-5,-5)$ is the set of points $(\pm 5, \pm 5)$. For any other starting point the sequence either converges to $(1,1)$, or does not converge at all. Some particular computed sequences are:

$(0.8,0.8),(1.09,1.09),(0.926,0.926),\ldots \rightarrow (1,1)$
$(3,3),(-1,-1),(1,1)$
$(-4,-4),(-2.75,-2.75),(-0.640,-0.640),(1.147,1.147),\ldots \rightarrow (1,1)$
$(6,6),(-7.75,-7.75),(-13.8,-13.8),\ldots$ does not converge
$(-5,-0.45),(1.12,-5),(-5,0.89),(1.05,-5),\ldots$ does not converge.

In the case when $T$ itself is not a contraction mapping, but some power of $T$ is, the method of successive substitutions still converges; this can be shown by modifying Theorem 5.2. The simplest version of this observation is given by the following theorem.

**THEOREM 5.3.** Let $T$ be an operator from a Banach space $X$ into itself and let $p$ be a positive integer such that

$$S = T^p \qquad (5.15)$$

is bounded with bound $\mu(S) < 1$. Then (5.2) has a unique fixed point $x^* \in X$ and the method of successive substitutions yields a sequence $\{x_n\}$ converging to $x^*$.

PROOF. Consider the sequence $\{x_n\}$ and look at the subsequence $\{z_i = x_{ip}\}$. Then

$$z_{i+1} = Sz_i$$

and since $S$ has bound less than one, $z_i \to z^* \in X$. Looking now at the subsequence $\{\bar{z}_i = x_{ip+1}\}$, we again have

$$\bar{z}_{i+1} = S\bar{z}_i$$

and $\bar{z}_i \to \bar{z}^* \in X$. But if $z^* \neq \bar{z}^*$, then

$$\|z^* - \bar{z}^*\| = \|Sz^* - S\bar{z}^*\| < \|z^* - \bar{z}^*\|.$$

Since this is a contradiction, we must have $z^* = \bar{z}^*$. The argument can be repeated for the other subsequences; hence $x_n \to x^*$, a unique fixed point.

∎

In a similar fashion the argument can be extended to apply when $S$ is a contraction mapping near $x^*$, but not on the whole of $X$. We leave this modification to the reader. Rall (1969) gives a somewhat different, but closely related, version of this result.

**Example 5.6.** Consider the ordinary differential equation

$$\frac{dx(t)}{dt} = f(t, x(t)) \qquad \text{in } [0, A], \tag{5.16}$$

$$x(0) = x_0,$$

where $f(t, x)$ is continuous in $t$ and satisfies the Lipschitz condition

$$|f(t, s_1) - f(t, s_2)| \leq M|s_1 - s_2|.$$

If we formally integrate (5.16) we get

$$x(t) = x_0 + \int_0^t f(s, x(s)) \, ds, \tag{5.17}$$

which is in the form of (5.2). If we take $X$ as the space $C[0, A]$ with the maximum norm, then

$$\|Tx_1 - Tx_2\| \leq \max \left| \int_0^t (f(s, x_1(s)) - f(s, x_2(s))) \, ds \right|$$

$$\leq MA \|x_1 - x_2\|.$$

If $MA < 1$ this is a contraction mapping and (5.17) has a unique solution. But we can do better:

$$T^2 x = x_0 + \int_0^t f\left(s, x_0 + \int_0^s f(\tau, x(\tau)) \, d\tau\right) ds,$$

so that

$$\|T^2 x_1 - T^2 x_2\| \leq \max\left|M \int_0^t Ms \, ds\right| \|x_1 - x_2\|$$

$$\leq \frac{M^2 A^2}{2} \|x_1 - x_2\|,$$

and, in general, for $p > 1$,

$$\|T^p x_1 - T^p x_2\| \leq \frac{M^p A^p}{p!} \|x_1 - x_2\|.$$

Thus for $p$ sufficiently large $T^p$ has bound less than one and, by Theorem 5.3, (5.17) has a unique solution. Since (5.16) and (5.17) are equivalent (show) it follows that the differential equation (5.16) has one and only one solution.

In most practical situations the iteration (5.6) cannot be carried out exactly. There may be discretization errors involved in computing $Tx_{n-1}$, but even if this is not the case there will always be some round-off error. Thus the actual iterates can be considered to be the result of the perturbed iteration

$$\hat{x}_n = T\hat{x}_{n-1} + \delta_n, \tag{5.18}$$

where $\hat{x}_0 = x_0$, and $\delta_n$ represents the total effect of all the errors at the $n$th stage. This will of course affect the final result and the contraction mapping theorem has to be modified accordingly.

**THEOREM 5.4.** Let $\hat{x}_n$ be computed by (5.18) and assume that the conditions of Theorem 5.2 are satisfied. If $\|\delta_n\| \leq \delta$, and $T$ is a contraction mapping with contraction factor $\theta < 1$ in $b(x_0, r)$ where

$$r \geq r_0 + \frac{\delta}{1 - \theta},$$

then

   (a)  $\hat{x}_n \in b(x_0, r)$.
   (b)  $\|\hat{x}_n - x^*\| \leq \theta^n r_0 + \dfrac{\delta}{1-\theta}$.

PROOF.   From (5.6) and (5.18)

$$\|x_n - \hat{x}_n\| \leq \|Tx_{n-1} - T\hat{x}_{n-1}\| + \|\delta_n\|.$$

Assume now that for $i = 1, 2, \ldots, n-1$ $\hat{x}_i \in b(x_0, r)$. Then

$$\begin{aligned}
\|x_n - \hat{x}_n\| &\leq \theta \|x_{n-1} - \hat{x}_{n-1}\| + \delta \\
&\leq \theta^2 \|x_{n-2} - \hat{x}_{n-2}\| + \theta\delta + \delta \\
&\quad\vdots \\
&\leq \theta^n \|x_0 - \hat{x}_0\| + (\theta^{n-1} + \theta^{n-2} + \cdots + 1)\delta.
\end{aligned}$$

Since $x_0 = \hat{x}_0$

$$\|x_n - \hat{x}_n\| \leq (\theta^{n-1} + \theta^{n-2} + \cdots + 1)\delta \leq \frac{\delta}{1-\theta}. \qquad (5.19)$$

Thus

$$\begin{aligned}
\|\hat{x}_n - x_0\| &\leq \|x_n - x_0\| + \|x_n - \hat{x}_n\| \\
&\leq r_0 + \frac{\delta}{1-\theta}, \qquad \text{from (5.12)},
\end{aligned}$$

and $\hat{x}_n \in b(x_0, r)$. Since $\hat{x}_1 \in b(x_0, r)$, Part (a) of the theorem follows by induction. The second part is obtained from

$$\|\hat{x}_n - x^*\| \leq \|\hat{x}_n - x_n\| + \|x_n - x^*\|,$$

and using (5.11) and (5.19) we get

$$\|\hat{x}_n - x^*\| \leq \frac{\delta}{1-\theta} + \theta^n r_0. \qquad \blacksquare$$

This indicates that the perturbed iterates no longer converge to $x^*$, but only to within $\delta/(1-\theta)$ of $x^*$. We can no longer talk of true convergence, but if $\delta$ is small this may not have much practical significance. We also note that if $\theta$ is near unity, the solution may be sensitive to small perturbations, so that the factor $1/(1-\theta)$ can be considered a condition number for the problem.

The contraction mapping theorem has proved to be a very powerful result in the theoretical study of nonlinear problems; for example, the arguments given in Example 5.6 provide an existence and uniqueness proof for the solution of certain ordinary differential equations. Computationally, the method of successive substitutions is very convenient, since it involves only the solution of a sequence of direct problems. But as a general approach the method has certain shortcomings. It is often difficult to find a formulation for which the method converges; even if one can be found the method is efficient only if $\theta$ is quite small. The method is thus useful only in special cases.

We are familiar with these difficulties in the root-finding case, where a number of "better" methods are known. One of these is the well-known *Newton's Method*: to find the root of $f(x)=0$ we compute a sequence of approximations by

$$x_{n+1} = x_n - \frac{f(x_n)}{f'(x_n)}. \tag{5.20}$$

Provided one starts with a good initial guess $x_0$ for the root, and $f' \neq 0$ at the root, this converges very quickly to the solution. In order to generalize (5.20) we must develop the notion of the derivative of an operator. In the next section we shall see how the results of calculus of functions can be extended to general operators.

### Exercises 5.1

1. Verify the statement made in Example 5.2.
2. For Example 5.3 (a) determine the accuracy of $z_3$ (without using the known solution), (b) show that the system has only one solution in all of $R^2$.
3. Show that the system

$$x + 4y^2 = 1.4$$
$$4x^2 + y = 1.6$$

has one and only one solution in $0 \leq x \leq 1$, $0 \leq y \leq 1$. Find the solution accurate to within 0.001, using the method of successive substitutions. Prove that the result you obtain is within the required tolerance.

4. Prove that the only solution of the nonlinear integral equation

$$x(t) + \sin t \int_0^1 \sin x(s)\, ds = 0, \qquad 0 \le t \le 1,$$

is $x(t) = 0$.

5. Modify Theorem 5.3 to apply when $S$ is a contraction mapping near $x^*$. Restate the theorem carefully and completely, then carry out the proof.

6. Consider the solution of the linear system

$$x + 0.9y + 0.1z = 1$$
$$0.4x + y + 0.4z = 0$$
$$0.8x + 0.1y + z = 0$$

by the iteration

$$\begin{pmatrix} x_n \\ y_n \\ z_n \end{pmatrix} = \begin{pmatrix} 1 \\ 0 \\ 0 \end{pmatrix} - \begin{pmatrix} 0 & 0.9 & 0.1 \\ 0.4 & 0 & 0.4 \\ 0.8 & 0.1 & 0 \end{pmatrix} \begin{pmatrix} x_{n-1} \\ y_{n-1} \\ z_{n-1} \end{pmatrix}.$$

Show that this iteration converges to the unique solution for arbitrary starting values.

7. Use the method of successive substitutions to solve

$$y'(t) - 0.1 \sin y(t) = 0$$
$$y(0) = 1, \qquad \text{in } 0 \le t \le 1,$$

to an accuracy of 0.01.

8. For the root-finding problem $t = g(t)$ show graphically why the problem is ill-conditioned if $1 - \theta$ is small.

9. Show that the integro-differential equation

$$x'(t) = x(t) + \int_0^t k(t,s)x(s)\, ds, \qquad 0 \le t \le T,$$
$$x(0) = x_0$$

with $|k(t,s)| \le K < \infty$, has a unique solution for all finite $T$.

10. Let $L_i$, $i = 1,\ldots,4$ be linear operators from a Banach space $X$ into itself. If $L_1$ and $L_4$ have bounded inverses on all of $X$ and $\|L_1^{-1}L_2\| < 1$,

$\|L_4^{-1}L_3\| < 1$, show that the operator system

$$L_1 x + L_2 y = z_1$$
$$L_3 x + L_4 y = z_2$$

has a unique solution $x \in X$, $y \in X$ for all $z_1, z_2 \in X$.

## 5.2  SOME ELEMENTARY RESULTS IN THE CALCULUS OF OPERATORS

We can introduce the idea of the derivative of an operator by generalizing the definition for the derivative of a function of one variable. Since such a generalization can be made in different ways we are led to several possible definitions for the derivative of an operator.

**DEFINITION 5.3.** Let $P: X \to Y$ be an operator between two normed linear spaces. If for $x_0 \in X$ there exists a linear operator $P_W'(x_0)$ such that for all $h \in X$

$$\lim_{s \to 0} \frac{P(x_0 + sh) - P(x_0)}{s} = P_W'(x_0)h, \qquad (5.21)$$

then we say that $P$ is *weakly differentiable* at $x_0$. The operator $P_W'(x_0)$ is called the *weak or Gateaux* derivative of $P$ at $x_0$.

For our purposes it will be sufficient and more convenient to use a somewhat more restrictive definition.

**DEFINITION 5.4.** Let $P: X \to Y$ be an operator between two normed linear spaces. If for $x_0 \in X$ there exists a *bounded* linear operator $P'(x_0)$ such that for all $\|\Delta x\| \to 0$

$$\lim_{\|\Delta x\| \to 0} \frac{\|P(x_0 + \Delta x) - P(x_0) - P'(x_0)\Delta x\|}{\|\Delta x\|} = 0, \qquad (5.22)$$

then we say that $P$ is *strongly differentiable* at $x_0$. The operator $P'(x_0)$ is called the *strong or Fréchet derivative*.

While there exists a close relation between the two notions they are not equivalent.

**THEOREM 5.5.**  If $P$ is strongly differentiable at $x_0$, then it is also weakly differentiable there and $P'_W(x_0) = P'(x_0)$.

PROOF.  We assign this as an exercise.  ∎

The converse of this theorem is not true; an operator may have a weak derivative but not a strong one. Here we are interested only in strong differentiability and we shall use the term derivative to mean Fréchet derivative.

It should be noted that while $P'(x_0)$ is a linear operator, it is generally not the same operator for different $x_0$; the notation $P'(x_0)$ is meant to indicate this dependence. We can also interpret $P'$ somewhat differently: for each $x_0 \in X$, $P'(x_0)$ is a bounded linear operator $X \to Y$. Thus we can think of $P'$ as defining a mapping $X \to L[X, Y]$; in this interpretation we think of $P'(x_0)$ as applying the operator $P'$ to $x_0$ yielding a linear operator. Since $P'$ is now itself an operator, we can again differentiate it, obtaining the second derivative of $P$, and so on. We look at this in more detail later.

**Example 5.7.**  Let $P: R^2 \to R^2$ be defined by

$$P\begin{pmatrix} x_1 \\ x_2 \end{pmatrix} = \begin{pmatrix} f_1(x_1, x_2) \\ f_2(x_1, x_2) \end{pmatrix}.$$

Then

$$P\begin{pmatrix} x_1 + \Delta x_1 \\ x_2 + \Delta x_2 \end{pmatrix} = \begin{pmatrix} f_1(x_1 + \Delta x_1, x_2 + \Delta x_2) \\ f_2(x_1 + \Delta x_1, x_2 + \Delta x_2) \end{pmatrix}$$

and by Taylor expansion (assuming $f_1$ and $f_2$ are sufficiently differentiable)

$$P\begin{pmatrix} x_1 + \Delta x_1 \\ x_2 + \Delta x_2 \end{pmatrix} = \begin{bmatrix} f_1(x_1, x_2) + \Delta x_1 \dfrac{\partial f_1}{\partial x_1} + \Delta x_2 \dfrac{\partial f_1}{\partial x_2} + \cdots \\ f_2(x_1, x_2) + \Delta x_1 \dfrac{\partial f_2}{\partial x_1} + \Delta x_2 \dfrac{\partial f_2}{\partial x_2} + \cdots \end{bmatrix}$$

where the derivatives are evaluated at $(x_1, x_2)$. If we take as $P'(x_1, x_2)$

$$L = \begin{bmatrix} \dfrac{\partial f_1}{\partial x_1} & \dfrac{\partial f_1}{\partial x_2} \\ \dfrac{\partial f_2}{\partial x_1} & \dfrac{\partial f_2}{\partial x_2} \end{bmatrix}$$

then it follows easily that (5.22) is satisfied (with any suitable norm), hence $L$ is the Fréchet-derivative of $P$.

In general, for $P: R^n \rightarrow R^m$ defined by

$$P \begin{bmatrix} x_1 \\ \vdots \\ x_n \end{bmatrix} = \begin{bmatrix} f_1(x_1, \ldots, x_n) \\ \vdots \\ f_m(x_1, \ldots, x_n) \end{bmatrix}$$

the derivative is the Jacobian matrix

$$P'(x) = \begin{bmatrix} \dfrac{\partial f_1}{\partial x_1} & \cdots & \dfrac{\partial f_1}{\partial x_n} \\ \vdots & & \vdots \\ \dfrac{\partial f_m}{\partial x_1} & \cdots & \dfrac{\partial f_m}{\partial x_n} \end{bmatrix}.$$

**Example 5.8.** If $L$ is a linear operator $X \rightarrow Y$, then for every $x_0 \in X$

$$L'(x_0) = L,$$

which is obvious from (5.22). This does not say that $L'$ and $L$ are identical, but that $L'$ has the same value $L$ at all points $x_0$. The observation is analogous to the fact that the derivative of a linear function is a constant.

**Example 5.9.** Let $X = C^{(1)}[0, 1]$, $Y = C[0, 1]$, both with maximum norm, and define $P$ by

$$Px = x^2 \frac{dx}{dt}.$$

Using Definition 5.3 we have

$$P(x_0 + sh) - P(x_0) = (x_0 + sh)^2 \frac{d}{dt}(x_0 + sh) - x_0^2 \frac{dx_0}{dt}$$

$$= s\left( x_0^2 \frac{dh}{dt} + 2hx_0 \frac{dx_0}{dt} \right) + O(s^2).$$

Hence the Gateaux derivative is

$$P_W'(x_0)[\quad] = x_0^2 \frac{d[\quad]}{dt} + 2x_0 \frac{dx_0}{dt}[\quad].$$

The rules for differentiation of functions familiar to us from introductory calculus apply also in this more general setting. In particular we have the following two useful results.

**THEOREM 5.6.**   If $P_1$ and $P_2$ are both differentiable at $x_0$, then

$$(\alpha P_1 + \beta P_2)'(x_0) = \alpha P_1'(x_0) + \beta P_2'(x_0) \tag{5.23}$$

where $\alpha$ and $\beta$ are any scalars.

PROOF.   This follows easily from Definition 5.4.        ∎

**THEOREM 5.7.**   If $P_1$ and $P_2$ are operators such that $P_2$ is differentiable at $x_0$ and $P_1$ is differentiable at $P_2(x_0)$, then

$$(P_1 P_2)'(x_0) = P_1'(P_2(x_0)) P_2'(x_0). \tag{5.24}$$

PROOF.   Note that, because of Definition 5.4, for any differentiable operator $P$

$$P(x_0 + \Delta x) = P(x_0) + P'(x_0)\Delta x + \eta_P(\Delta x),$$

where

$$\lim_{\|\Delta x\| \to 0} \|\eta_P(\Delta x)\| / \|\Delta x\| = 0.$$

Then

$$
\begin{aligned}
P_1 P_2(x_0 + \Delta x) &= P_1(P_2(x_0 + \Delta x)) \\
&= P_1\big[ P_2(x_0) + P_2'(x_0)\Delta x + \eta_{P_2}(\Delta x) \big] \\
&= P_1 P_2(x_0) + P_1'(P_2(x_0)) P_2'(x_0)\Delta x \\
&\quad + P_1'(P_2(x_0))\eta_{P_2}(\Delta x) + \eta_{P_1}\big(P_2'(x_0)\Delta x + \eta_{P_2}(\Delta x)\big),
\end{aligned}
$$

and hence

$$\lim_{\|\Delta x\| \to 0} \frac{\| P_1 P_2(x_0 + \Delta x) - P_1 P_2(x_0) - P_1'(P_2(x_0)) P_2'(x_0)\Delta x \|}{\|\Delta x\|} = 0,$$

which proves (5.24). This differentiation rule is the generalization of the chain rule in ordinary calculus. It should, of course, be kept in mind that the product of two operators corresponds to the composition of two functions and not their usual product.                                          ∎

We can also define the integral of an operator. Let $P:X\to Y$, and let $x_1$ and $x_2$ be two distinct points in $X$. The *line segment* between $x_1$ and $x_2$ is defined as the set of all points $x$ such that

$$x = tx_1 + (1-t)x_2, \qquad 0 \le t \le 1.$$

Let $\pi$ be a partition of $[0,1]$ by the points $0 = t_0 < t_1 < \cdots < t_n = 1$, and introduce the points $t_i^*$ such that $t_i \le t_i^* \le t_{i+1}$. We then consider

$$S_n = \sum_{i=0}^{n-1} P(t_i^*)(t_{i+1} - t_i), \qquad (5.25)$$

where $P(t):R^1\to Y$ is defined by $P(t) = P(tx_1 + (1-t)x_2)$.

**DEFINITION 5.5.** If the sequence $S_n$ defined by (5.25) converges to $S \in Y$ as $n\to\infty$ and $|\pi| = \max|t_{i+1} - t_i| \to 0$, then $P$ is said to be integrable and $S$ is the definite integral of $P$ between the limits 0 and 1 along the line segment between $x_1$ and $x_2$. We write this as

$$S = \int_0^1 P(t)\,dt. \qquad (5.26)$$

This definition is a generalization of the usual definition of an integral by Riemann sums. As we have no further use for integrals we omit any details. A more thorough discussion can be found in Rall (1969, pp. 120–127).

If $X$ and $Y$ are linear spaces then $L[X,Y]$ is itself a linear space. Consider now the set of all bounded linear operators $X\to L[X,Y]$. This set, which we denote by $L[X,L[X,Y]]$ or $L[X^2,Y]$ is again a linear space. An element of $L[X^2,Y]$ is said to be a *bilinear operator*. We can extend the process and consider the set of all linear operators $X\to L[X^2,Y]$, and so on, and we have

**DEFINITION 5.6.** For $k = 1, 2, \ldots$, let $L[X^{k+1}, Y]$ denote the space of all bounded linear operators $X\to L[X^k, Y]$, with $L[X^1, Y] = L[X, Y]$. An element of $L[X^k, Y]$ is called a *k-linear operator*.

**Example 5.10.** If $X = Y = R^2$, then $L[X,Y]$ is the set of all $2\times2$ matrices. The bilinear operators are the operations which transform vectors into matrices; they can be represented by $2 \times 2 \times 2$ arrays. Thus if the bilinear operator $B$ has elements $b_{ijk}$, $i,j,k = 1,2$, then

$$(Bx)_{ij} = \sum_{k=1}^{2} b_{ijk} x_k$$

defines a $2 \times 2$ matrix.

If $A_k$ is a $k$-linear operator, then $A_k x_k$ is a $(k-1)$-linear operator and $(\ldots(A_k x_k) x_{k-1}) \ldots) x_1$ is an element of $Y$. Thus $A_k$ can also be considered as an operator from the product space $X \times X \times \cdots \times X \rightarrow Y$, defined by

$$A_k(x_k, x_{k-1}, \ldots, x_1) = A_k x_k x_{k-1} \ldots x_1$$
$$= (\ldots((A_k x_k) x_{k-1}) \ldots) x_1.$$

For notational convenience we use the short-hand notation

$$A_k x^k = A_k x x \ldots x.$$

**DEFINITION 5.7.** If for $k = 0, 1, \ldots, n$ $A_k$ is a $k$-linear operator, then

$$A_n x^n + A_{n-1} x^{n-1} + \cdots + A_1 x + A_0$$

is a *generalized polynomial* of degree $n$. ($A_0$ is an element of $Y$.)

The norm of a $k$-linear operator $A_k : X \rightarrow L[X^{k-1}, Y]$ is, by the usual definition,

$$\|A_k\| = \sup_{\|x\|=1} \|A_k x\|_{L[X^{k-1}, Y]}.$$

We can get this into a more manageable form by noting that this is equivalent to

$$\|A_k\| = \sup_{\|x_1\|=1, \ldots, \|x_k\|=1} \|A_k x_k x_{k-1} \ldots x_1\|. \tag{5.27}$$

We will leave the proof of this to the reader.

Since, as already remarked, we can consider $P'$ as an operator $X \rightarrow L[X, Y]$ we can differentiate it by applying Definition 5.4. Thus, if there exists a bilinear operator $B$ such that

$$\lim_{\|\Delta x\| \to 0} \frac{\|P'(x_0 + \Delta x) - P'(x_0) - B \Delta x\|}{\|\Delta x\|} = 0 \tag{5.28}$$

then $B$ is the second derivative of $P$ at $x_0$, denoted by $P''(x_0)$. In general, we have Definition 5.8.

**DEFINITION 5.8.** If $P$ is differentiable $k-1$ times in some neighborhood $b(x_0, r)$, $r > 0$, of $x_0$, and if a $k$-linear operator $K$ exists such that

$$\lim_{\|\Delta x\| \to 0} \frac{\| P^{(k-1)}(x_0 + \Delta x) - P^{(k-1)}(x_0) - K\Delta x \|}{\|\Delta x\|} = 0, \qquad (5.29)$$

then $K = P^{(k)}(x_0)$ is the $k$th derivative of $P$ at $x_0$.

**Example 5.11.** Consider the operator $P$ defined in Example 5.7. Here

$$P'(\mathbf{x}_0) = \begin{bmatrix} \dfrac{\partial f_1}{\partial x_1} & \dfrac{\partial f_1}{\partial x_2} \\[3mm] \dfrac{\partial f_2}{\partial x_1} & \dfrac{\partial f_2}{\partial x_2} \end{bmatrix}_{x = x_0} .$$

so that

$$P'(\mathbf{x}_0 + \Delta \mathbf{x}) - P'(\mathbf{x}_0) =$$

$$\begin{bmatrix} \dfrac{\partial^2 f_1}{\partial^2 x_1}\Delta x_1 + \dfrac{\partial^2 f_1}{\partial x_1 \partial x_2}\Delta x_2 & \dfrac{\partial^2 f_1}{\partial x_1 \partial x_2}\Delta x_1 + \dfrac{\partial^2 f_1}{\partial^2 x_2}\Delta x_2 \\[4mm] \dfrac{\partial^2 f_2}{\partial^2 x_1}\Delta x_1 + \dfrac{\partial^2 f_2}{\partial x_1 \partial x_2}\Delta x_2 & \dfrac{\partial^2 f_2}{\partial x_1 \partial x_2}\Delta x_1 + \dfrac{\partial^2 f_2}{\partial^2 x_2}\Delta x_2 \end{bmatrix}_{x = x_0}$$

$$+ \text{higher order terms.}$$

The second derivative at $P$ at $x_0$ is then the $2 \times 2 \times 2$ array $B$ with elements

$$b_{ijk} = \frac{\partial^2 f_i}{\partial x_j \partial x_k}\bigg)_{x = x_0} .$$

**Example 5.12.** If $L$ is a linear operator, then its second derivative is zero. We already know, from Example 5.8, that

$$L'(x_0) = L.$$

Therefore

$$L'(x_0 + \Delta x) - L'(x_0) = L - L = 0,$$

and hence $L'' = 0$.

**Example 5.13.**   Let $K$ be the integral operator $C[0,1] \rightarrow C[0,1]$ defined by

$$Kx(s) = \int_0^1 k(s,t)x^2(t)\,dt.$$

Then

$$K'(x_0)[\quad](s) = 2\int_0^1 k(s,t)x_0(t)[\quad]\,dt,$$

$$K''(x_0)[\quad][\quad](s) = 2\int_0^1 k(s,t)[\quad][\quad]\,dt,$$

$$K'''(x_0) = 0,$$

as is easily verified.

In elementary numerical analysis Taylor's theorem is frequently used in analyzing algorithms. For the general operator case there exists an extension of Taylor's theorem which is equally useful.

**THEOREM 5.8.   GENERALIZED TAYLOR'S THEOREM.**   Let $P : X \rightarrow Y$ be an operator between two Banach spaces, such that $P$ is $n$ times continuously differentiable in a neighborhood $b(x_0, r)$, $r > 0$, of $x_0$. Then for all $x$ in the interior of $b(x_0, r)$

$$\left\| Px - Px_0 - \sum_{i=1}^{n-1} \frac{1}{i!} P^{(i)}(x_0)(x-x_0)^i \right\| \leq \sup_{\bar{x} \in l(x_0, x)} \| P^{(n)}(\bar{x})\| \frac{\|x-x_0\|^n}{n!}$$

(5.30)

where $l(x_0, x)$ is the line segment between $x_0$ and $x$.

PROOF.   A proof of this theorem, under somewhat less restrictive conditions, can be found in Rall (1969, pp. 124–126).   ∎

The material presented in this section has been adapted, in abbreviated form, from Rall (1969). More details can be found there, as well as in (Kantorovich and Akilov, 1964, Ch. 17).

### Exercises 5.2

1. Verify the expressions for the derivative in Example 5.13.

2. Compute the first derivative at $x_0$ of the following operators.

$$\text{(a)}\quad Px = \frac{dx(s)}{ds} + e^{-x(s)}.$$

$$\text{(b)}\quad Px = \int_0^1 k(s,t)x^3(t)\,dt.$$

3. Show that if $\|T'(x)\| < 1$ for $x$ in some neighborhood of $x_0$, then $T$ is a contraction mapping near $x_0$.

4. Show that if $P$ is differentiable at $x_0$, then it is also continuous there.

5. If $R = LP$, where $L$ is a linear operator show that

$$R' = LP'.$$

6. Let $P$ be an operator from $X$ into the product space $Y_1 \times Y_2$ defined by

$$Px = (P_1x, P_2x).$$

If $P_1$ and $P_2$ are both differentiable at $x_0$ show that

$$P'(x_0)[\ \ ] = (P_1'(x_0)[\ \ ], P_2'(x_0)[\ \ ]).$$

7. Let $P: L[X] \to L[X]$ be defined by $P(L) = L^2$. Find $P'(L_0)$.

8. For $P: R^2 \to R^1$ defined by $P(x,y) = x^2 + y^4$ find $P'(x_0)$ and $P''(x_0)$.

9. If $P$ is continuously differentiable, show that

$$P(1) - P(0) = \int_0^1 P'(t)\,dt.$$

10. Find the second and third derivatives of the operators in Exercise 2.

11. (a)   If $A_2$ is a bilinear operator show that

$$\|A_2 x_1 x_2\| \le \|A_2\|\,\|x_1\|\,\|x_2\|.$$

(b)   Prove that $\|A_2\| = \sup\limits_{\|x_1\| = 1,\,\|x_2\| = 1} \|A_2 x_1 x_2\|.$

(c)   Prove (5.27) for arbitrary $k$.

12. If $X$ and $Y$ are Banach spaces, show that $L(X^k, Y)$, $k = 1, 2, \ldots$ are also Banach spaces.

## 5.3   THE GENERALIZED NEWTON'S METHOD

We are now ready to extend (5.20) to the general operator case. In order to evaluate the procedure we so obtain we must compare it to the method of successive substitutions. One criterion for such a comparison is the order of convergence defined by Definition 5.9.

**DEFINITION 5.9.** Let $x_0, x_1, x_2, \ldots$ be a sequence converging to $x^*$. If there exist constants $k$ and $p$ such that

$$\|x_{n+1} - x^*\| \leq k\|x_n - x^*\|^p, \tag{5.31}$$

then the sequence is said to converge to $x^*$ with order $p$. A method which produces such a sequence is said to have an order of convergence $p$.

Note that this definition of order of convergence differs from that used in the analysis of discretization methods (Section 4.3). There is no contradiction, since we are talking about different types of procedures; the $n$ here is not a discretization parameter, but the step number in an iterative process.

We can derive the formula for the generalized Newton's method by a simple, intuitive argument. Let $x^*$ be a solution to $P(x) = 0$; then for $x$ near $x^*$

$$0 = P(x^*) = P(x) + P'(x)(x^* - x) + \eta(x, x^*),$$

where $\eta(x, x^*)$ is small. If we neglect $\eta(x, x^*)$ and solve what is left for $x^*$, we do not obtain the exact solution, but hopefully a better approximation to it. This suggests the iterative process

$$P'(x_n)x_{n+1} = P'(x_n)x_n - P(x_n), \tag{5.32}$$

or in alternate form

$$x_{n+1} = x_n - [P'(x_n)]^{-1}P(x_n), \tag{5.33}$$

which is a generalization of (5.20) for operator equations. For computational purposes it is convenient to rewrite this as

$$x_{n+1} = x_n + \Delta x_n, \tag{5.34}$$

where $\Delta x_n$ is the solution of

$$P'(x_n)\Delta x_n = -P(x_n). \tag{5.35}$$

Since $P'(x_n)$ is a linear operator, computing the correction $\Delta x_n$ requires the solution of a linear equation. Thus Newton's method is an iterative technique in which each step involves the solution of a linear problem.

These arguments make the technique plausible; more complete informa-tion is needed to analyze its effectiveness.

**THEOREM 5.9.** Let $P:X \to Y$ be an operator between two Banach spaces and assume that $P(x)$, $[P'(x)]^{-1}$, and $P''(x)$ are bounded in some neighborhood $b(x^*, r)$, with $P(x^*) = 0$. Then Newton's method has second-order convergence near $x^*$.

**PROOF.**

$$x_{n+1} - x^* = x_n - [P'(x_n)]^{-1}P(x_n) - x^*$$
$$= -[P'(x_n)]^{-1}[P(x_n) - P'(x_n)(x_n - x^*)],$$

so that

$$\|x_{n+1} - x^*\| \le \|[P'(x_n)]^{-1}\| \, \|P(x_n) - P'(x_n)(x_n - x^*)\|$$
$$= \|[P'(x_n)]^{-1}\| \, \|P(x^*) - P(x_n) - P'(x_n)(x^* - x_n)\|.$$

From the generalized Taylor's theorem (5.30), with $n = 2$, $x = x^*$, $x_0 = x_n$, we then have

$$\|x_{n+1} - x^*\| \le \|[P'(x_n)]^{-1}\| \sup_{x \in b(x^*, r)} \|P''(x)\| \frac{\|x_n - x^*\|^2}{2}. \qquad \blacksquare$$

This shows that the Newton's method has second-order convergence, provided $x_0$ is sufficiently close to $x^*$ so that successive iterates stay in $b(x^*, r)$. A more precise result is provided by a theorem due to *Kantorovich*.

**THEOREM 5.10.** Assume that the conditions of Theorem 5.9 are satis-fied and that there exist constants $B$, $\eta$, $k$, and

$$h = B\eta k \le \frac{1}{2},$$
$$r \ge (1 - \sqrt{1 - 2h})\eta/h,$$

such that

$$\|P'(x_0)^{-1}\| \le B,$$
$$\|x_1 - x_0\| \le \eta,$$
$$\|P''(x)\| \le k, \qquad \text{for all } x \in b(x_0, r).$$

Then Newton's method, starting with $x_0$, converges to $x^*$ and

$$\|x^* - x_n\| \le \frac{(2h)^{2^n - 1}\eta}{2^{n-1}}. \tag{5.36}$$

The proof of this theorem is somewhat complicated and will be omitted; the interested reader may consult (Kantorovich and Akilov, 1964, Ch. 18) and (Rall, 1969, pp. 135–138).

While the Kantorovich theorem gives sufficient conditions under which the method converges, it is not always easy to use since the conditions are difficult to check. From a practical viewpoint the information in Theorem 5.9 is usually sufficient. Starting with some initial guess $x_0$ one applies Newton's method; if it converges it usually does so very quickly, if not, one must find a better initial approximation.

**Example 5.14.** Find a solution to

$$Px = P\begin{bmatrix} x \\ y \end{bmatrix} = \begin{bmatrix} x + 4y^2 - \frac{3}{2} \\ 4x^2 + y - \frac{3}{2} \end{bmatrix} = 0.$$

Here

$$P'(x) = \begin{pmatrix} 1 & 8y \\ 8x & 1 \end{pmatrix}.$$

With

$$x_0 = \begin{pmatrix} 0.4 \\ 0.4 \end{pmatrix},$$

the first correction $\Delta x_0$, computed by

$$P'(x_0)\Delta x_0 = -P(x_0)$$

is

$$\Delta x_0 = \begin{pmatrix} 0.1095238 \\ 0.1095238 \end{pmatrix}.$$

Continuing the process, we get

$$x_1 = \begin{pmatrix} 0.5095238 \\ 0.5095238 \end{pmatrix}, \quad x_2 = \begin{pmatrix} 0.5000714 \\ 0.5000714 \end{pmatrix},$$

and the next iterate $x_3$ agrees with the exact solution $(0.5, 0.5)$ to an accuracy better than $5 \times 10^{-9}$. As expected the convergence is very fast.

**Example 5.15.**   For the nonlinear system

$$P \begin{bmatrix} x_1 \\ \vdots \\ x_n \end{bmatrix} = \begin{bmatrix} f_1(x_1, \ldots, x_n) \\ \vdots \quad \vdots \\ f_n(x_1, \ldots, x_n) \end{bmatrix} = 0$$

we know from Examples 5.7 and 5.11, that $P'(x)$ is an $n \times n$ matrix with elements $\partial f_i / \partial x_j$ and $P''(x)$ is an $n \times n \times n$ array with elements $\partial^2 f_i / \partial x_j \partial x_k$. If we use the maximum norm for $x$ then

$$\|P'(x)\| = \max_i \sum_{j=1}^{n} \left| \frac{\partial f_i}{\partial x_j} \right|.$$

Also, from (5.27)

$$\|P''(x)\| = \sup_{\|y\|=1, \|z\|=1} \max_i \sum_{j,k} \left| \frac{\partial^2 f_i}{\partial x_j \partial x_k} y_j z_k \right|$$

so that

$$\|P''(x)\| \leq \max_i \sum_{j,k} \left| \frac{\partial^2 f_i}{\partial x_j \partial x_k} \right|.$$

To use the Kantorovich theorem we need to be able to bound these partial derivatives near the solution.

**Example 5.16.**   Consider the nonlinear integral equation

$$Px(s) = x(s) + \int_0^1 K(s,t) x^2(t) \, dt - y(s) = 0.$$

Then

$$P'(x_0)[\quad] = I[\quad] + 2 \int_0^1 K(s,t) x_0(t)[\quad] \, dt.$$

Using (5.35) we see that Newton's method for this problem has the form

$$x_{n+1}(s) = x_n(s) + \Delta x_n(s),$$

where $\Delta x_n(s)$ is the solution of the linear integral equation

$$\Delta x_n(s) + 2\int_0^1 K(s,t)x_n(t)\Delta x_n(t)\,dt = y(s) - x_n(s) - \int_0^1 K(s,t)x_n^2(t)\,dt.$$

In actual applications of Newton's method several problems arise which merit closer attention.

(a)  One of the most serious problems lies in finding a sufficiently close initial approximation. In general, this is quite difficult and no universally effective scheme is known. Sometimes the physical situation from which the equation arises will give some information; hopefully this will be good enough to obtain convergence.

(b)  Each step of the method may be quite time-consuming since one needs to evaluate $P'(x_n)$ and then to solve a linear equation. One can sometimes achieve better results by not recomputing $P'(x_n)$ at each step, using $P'(x_0)$ instead. Thus we obtain the *modified Newton's method*

$$\bar{x}_{n+1} = \bar{x}_n - [P'(x_0)]^{-1}P(\bar{x}_n). \tag{5.37}$$

With this we can often eliminate a considerable part of the computational work at each step. Of course, as one might suspect, the rate of convergence suffers; in fact, the order of convergence is reduced to one. It can be shown that if the conditions of Theorem 5.10 are satisfied, then

$$\|x^* - \bar{x}_n\| \le 2h\eta(1-\sqrt{1-2h}\,)^{n-1}. \tag{5.38}$$

The proof of this, which is somewhat lengthy and technical, is given in (Rall, 1969, pp. 199–200).

When $h \ll \frac{1}{2}$, $\bar{x}_n$ converges to $x^*$ quite rapidly. If $h$ is near $\frac{1}{2}$, further efficiency can sometimes be achieved by alternating one step of Newton's method with several steps of the modified Newton's method.

Both methods considered so far, the method of successive substitutions and Newton's method, reduce the solution of the nonlinear problem to a sequence of simpler problems, that is, a sequence of direct problems or a sequence of linear inverse problems. One can of course attempt to approach this problem more directly by generalizing the methods developed in Chapter 4 to the nonlinear case. Let us, for example, reconsider the minimum residual methods of Section 4.2. To solve $P(x)=0$ we approximate the solution $x^*$ by a finite linear combination of some expansion functions $\varphi_{ni}$

$$x^* \simeq x_n(\alpha) = \sum_{i=1}^{n} \alpha_i \varphi_{ni}.$$

If we write

$$J(\alpha) = \| P(x_n(\alpha)) \| = \left\| P\left( \sum_{i=1}^{n} \alpha_i \varphi_{ni} \right) \right\|,$$

then clearly the functional $J(\alpha) \geq 0$, and $J(\alpha) = 0$ only if $x_n(\alpha)$ is a solution of the equation. In general, it will of course not be possible to find an $\alpha$ such that $J(\alpha) = 0$ exactly, but as in Section 4.2 we can try to find $\alpha$ such that $J(\alpha)$ is minimized. This leads us to the problem of *minimization of functional* which we will now consider briefly.

### Exercises 5.3

1. Prove the following result: Let $x^*$ be a fixed point of $T$ and assume that $T$ is $p$ times continuously differentiable in some neighborhood $b(x^*, r)$ of $x^*$. If $T'(x^*) = T''(x^*) = T^{(p-1)}(x^*) = 0$, but $T^{(p)}(x^*) \neq 0$, then the order of convergence of the method of successive substitutions is $p$, provided $x_0$ is sufficiently close to $x^*$. More specifically, show that

$$\| x_n - x^* \| \leq k_n \| x_0 - x^* \|^{p^n},$$

with $k_n = k^{(p^n - 1)/(p-1)}$, where $k$ is a bound for $\| T^{(p)} \| / p!$ in $b(x^*, r)$.

2. Starting with $x_0 = 0.8$, $y_0 = 0.8$ carry out two steps of Newton's method for

$$x^2 + y^2 = 2.1$$
$$3x^2 - y^2 = 2.0.$$

Prove that the process converges and find a bound for the error in the second iterate.

3. What effect does a small perturbation have on Newton's method? Specifically, if $\hat{x}_n$ is determined by

$$\hat{x}_{n+1} = \hat{x}_n + \Delta \hat{x}_n, \quad \hat{x}_0 = x_0,$$
$$P'(\hat{x}_n)\Delta \hat{x}_n = -P(\hat{x}_n) + \delta_n,$$

find a bound for $\| x_n - \hat{x}_n \|$.

4. If $B$ is a bilinear operator on $R^n$ with elements $b_{ijk}$, then, as already

shown in Example 5.15, using the maximum norm on $R^n$,

$$\|B\| \leq \max_i \sum_{j,k} |b_{ijk}|.$$

Can one claim that, in general,

$$\|B\| = \max_i \sum_{j,k} |b_{ijk}|?$$

5. Give a graphical interpretation of the modified Newton's method for the rootfinding problem $f(x) = 0$.

## 5.4   THE MINIMIZATION OF FUNCTIONALS

The study of the minimization of functionals, or optimization theory, is one of the major topics of computational mathematics. Even a summary of the most important results of this theory would require considerable space. Here we will present only some of the most basic ideas to acquaint the reader with the subject; for a more complete account one should consult some of the many texts on this topic. For a discussion of the theoretical foundations we mention in particular the books by Daniel (1971) and Céa (1971).

To simplify our discussion as much as possible we restrict the setting of the problem through the following assumptions:

$X$ is a Hilbert space.

$J : X \to R^1$ is a functional defined on all of $X$, such that $J'(x), J''(x), J'''(x)$ exist and are bounded for all $x \in X$.

$J$ is bounded below, that is $J(x) \geq l > -\infty$ for all $x \in X$.

There are two different situations which arise in practice: *unconstrained* minimization, where we look for a minimum of $J(x)$ for all possible $x$, and *constrained* minimization, where $x$ is required to lie in some subset of $X$ defined by imposing some conditions (constraints) on $x$. We will consider only the simpler case of unconstrained minimization.

**DEFINITION 5.10.**   $x^*$ is called a *global* minimum of $J$ if

$$J(x^*) \leq J(x), \qquad \text{for all } x \in X.$$

$x^*$ is called a *local* minimum of $J$ if

$$J(x^*) \leq J(x) \qquad \text{for all } x \text{ in some neighborhood } b(x^*, r)$$
of $x^*$.

As in elementary calculus there are some simple conditions which characterize a local minimum.

**THEOREM 5.11.**  If

$$J'(x^*) = 0 \qquad\qquad (5.39)$$

and $J''(x^*)$ is positive definite, by which we mean that there exists a $\gamma > 0$ such that

$$J''(x^*)x^2 \geq \gamma\|x\|^2, \qquad \text{for all } x \in X,$$

then $x^*$ is a local minimum of $J$.

PROOF.   Let $x$ be a point near $x^*$. Then, from (5.30),

$$J(x) - J(x^*) - J'(x^*)(x-x^*) = \frac{1}{2}J''(x^*)(x-x^*)^2 + \eta(x,x^*),$$

where

$$\lim_{x \to x^*} \frac{\eta(x,x^*)}{\|x-x^*\|^2} = 0.$$

Hence, for $\|x - x^*\|$ sufficiently small

$$\frac{J(x)-J(x^*)}{\|x-x^*\|^2} \geq \frac{1}{2}\gamma - \frac{|\eta(x,x^*)|}{\|x-x^*\|^2} \geq 0. \qquad \blacksquare$$

**THEOREM 5.12.**   If $x^*$ is a local minimum of $J$, then

$$J'(x^*) = 0.$$

PROOF.   Assume that $J'(x^*) \neq 0$. Since $J'(x^*)$ is linear we can always find an $x$, with $\|x - x^*\|$ as small as desired, such that

$$J'(x^*)(x-x^*) < -\gamma\|x - x^*\|$$

for some $\gamma > 0$ (Verify). Then, from (5.30)

$$J(x) - J(x^*) < -\gamma\|x - x^*\| + \eta(x,x^*),$$

where

$$\lim_{x \to x^*} \frac{\eta(x, x^*)}{\|x - x^*\|} = 0.$$

Thus there is some $x \neq x^*$ such that

$$J(x) < J(x^*)$$

contradicting the assumption that $x^*$ is a local minimum and hence the theorem is proved. It is, however, not necessary that $J''(x^*)$ be positive definite, as can be seen from the simple example $J(x) = x^4$, $x \in R^1$. ∎

There are several other simple properties that are useful in characterizing minima.

**DEFINITION 5.11.** A functional $J(x)$ is said to be *convex* in some subset $S$ of $X$ if for any $x_1, x_2 \in S$

$$J(\theta x_1 + (1 - \theta)x_2) \leq \theta J(x_1) + (1 - \theta)J(x_2), \qquad \text{for all } 0 < \theta < 1.$$

$$(5.40)$$

If strict inequality holds in (5.40) then $J$ is said to be *strictly* convex.

**THEOREM 5.13.** If $J$ is strictly convex in $S$, then there exists at most one local minimum in $S$.

**PROOF.** Assume that there are two distinct local minima $x^*$ and $x^{**}$ in $S$. Let $z$ be a point on $\hat{l}(x^*, x^{**})$, the interior of the line segment $l(x^*, x^{**})$, that is,

$$z = x^* + (1 - \theta)x^{**}, \qquad 0 < \theta < 1.$$

Then, from strict convexity,

$$J(z) < \theta J(x^*) + (1 - \theta)J(x^{**}).$$

Without loss of generality we may assume that $J(x^*) \geq J(x^{**})$, so that

$$J(z) < J(x^*).$$

But since $x^*$ is a local minimum there exists a neighborhood $b(x^*, r)$ such that $J(x^*) \leq J(z)$ for all $z \in b(x^*, r)$. By picking $z \in b(x^*, r) \cap \hat{l}(x^*, x^{**})$ we then obtain a contradiction, and the theorem is proved. ∎

From the above it also follows that if $J$ is strictly convex in all of $X$ then any local minimum is a global minimum. If $J$ is convex but not strictly so, then the minimum is not necessarily unique (Exercise 3).

Most algorithms for minimization are based on the same simple idea: starting with some $x_0$ we generate a sequence

$$x_{n+1} = x_n + \Delta x_n,$$

where $\Delta x_n$ is to be chosen such that

$$J(x_{n+1}) \leq J(x_n).$$

The way in which the $\Delta x_n$ are chosen determines whether the sequence converges to a minimum as well as the speed of convergence.

**DEFINITION 5.12.**   If $x^*$ is a local minimum of $J$ and $\{x_n\}$ is a sequence such that

$$J(x_n) \rightarrow J(x^*),$$

then $\{x_n\}$ is called a *minimal* sequence. If $\{x_n\}$ is such that

$$\lim_{n\to\infty} \|J'(x_n)\| = 0,$$

then $\{x_n\}$ is said to be a *critical* sequence.

A critical sequence is of course not necessarily minimal, nor are minimal or critical sequences necessarily convergent. A major concern in minimization is the problem of finding methods which will yield convergent minimal sequences. Often one can show more easily that a certain algorithm gives a critical sequence and one must then use additional information about $J$ (e.g., convexity, positive-definiteness of $J''$) to determine whether the sequence converges to a minimum.

Before proceeding we need to introduce an important theorem from functional analysis.

**THEOREM 5.14.   RIESZ REPRESENTATION THEOREM.**   Let $X$ be a Hilbert space and $f$ be any element in $X^*$. Then there exists a unique $h \in X$ such that

$$f(x) = (h, x), \qquad \text{for all } x \in X.$$

PROOF.  See any book on functional analysis, for example (Taylor, 1964, p. 245). ∎

**DEFINITION 5.13.**  The *gradient* of $J$ at $x_0$, denoted by $\nabla J(x_0)$, is an element of $X$ such that

$$J'(x_0)x = (\nabla J(x_0), x), \qquad \text{for all } x \in X. \qquad (5.41)$$

Since $J'(x_0)$ is a bounded linear functional, $\nabla J(x_0) \in X$ is uniquely defined by the Riesz representation theorem. It should be noted, however, that $J'(x_0)$ and $\nabla J(x_0)$ are not identical, since the first is a linear functional and the second an element of $X$. But since they are "equivalent" in the sense of (5.41) some authors tend to use the two terms without making a careful distinction.

**Example 5.17.**  Let $J : R^2 \rightarrow R^1$ be defined by

$$J(x,y) = f(x,y).$$

With the usual inner product $(z_1, z_2) = z_1^T z_2$ we have

$$J'(x_0, y_0) = \big( f_x(x_0, y_0), f_y(x_0, y_0) \big)$$

and

$$\nabla J(x_0, y_0) = \begin{pmatrix} f_x(x_0, y_0) \\ f_y(x_0, y_0) \end{pmatrix}.$$

If at a given point $x_n$ the gradient is not zero, then we can always find a $\Delta x_n$ such that

$$(\nabla J(x_n), \Delta x_n) < 0,$$

that is, such that $J$ decreases instantaneously in the direction $\Delta x_n$. We can therefore consider generating $\{x_n\}$ by the iterative process

$$x_{n+1} = x_n - s_n w_n, \qquad (5.42)$$

where

$$\begin{aligned} \|w_n\| &= 1, \\ (\nabla J(x_n), w_n) &\geq 0, \qquad (5.43) \\ s_n &> 0. \end{aligned}$$

The unit vector $w_n$ defines a direction of decreasing $J$ (direction of descent) and $s_n$ determines the step-size. If $w_n$ and $s_n$ are chosen properly the process (5.42) can be made to converge.

**THEOREM 5.15.** Assume that $\|J''(x)\| \leq K$ for all $x \in X$, and assume that there exist positive constants $k_1, k_2, \delta < 2$, such that the following are satisfied:

$$\gamma_n = (\nabla J(x_n), w_n) \geq k_1 \| \nabla J(x_n) \|, \tag{5.44}$$

$$k_2 \| \nabla J(x_n) \| \leq s_n \leq \frac{(2-\delta)k_1 \| \nabla J(x_n) \|}{K}. \tag{5.45}$$

Then the $x_n$ computed by (5.42) form a critical sequence.

PROOF.   From the generalized Taylor's theorem

$$J(x_{n+1}) - J(x_n) = -s_n(\nabla J(x_n), w_n) + \eta_n,$$

where

$$|\eta_n| \leq \frac{1}{2} s_n^2 K.$$

Now

$$\frac{|\eta_n|}{s_n \gamma_n} \leq \frac{s_n K}{2\gamma_n} \leq \frac{(2-\delta)k_1 \| \nabla J(x_n) \|}{2\gamma_n}, \quad \text{from (5.45),}$$

$$\leq 1 - \frac{1}{2}\delta, \quad\quad\quad\quad \text{from (5.44).}$$

Since

$$J(x_{n+1}) - J(x_n) = -s_n \gamma_n \left( 1 - \frac{\eta_n}{s_n \gamma_n} \right)$$

it follows that $\{J(x_n)\}$ is a nonincreasing sequence and

$$|J(x_{n+1}) - J(x_n)| \geq s_n \gamma_n \delta / 2.$$

Assume now that $\{x_n\}$ is not a critical sequence, that is, $\|\nabla J(x_n)\| \nrightarrow 0$. Then there exists an infinite subsequence $\{x_{n_j}\}$ of $\{x_n\}$ such that

$$\| \nabla J(x_{n_j}) \| \geq \mu > 0.$$

Thus $\gamma_{n_j} \geq k_1 \mu$, $s_{n_j} \geq k_2 \mu$, and

$$|J(x_{n_{j+1}}) - J(x_{n_j})| \geq \mu^2 k_1 k_2 \delta/2 > 0.$$

This implies that $\lim J(x_{n_j}) = -\infty$, which contradicts our assumption that $J$ is bounded below. Therefore $\{x_n\}$ must be a critical sequence. ■

It is instructive to consider an intuitive explanation of the various conditions we have imposed. Equation (5.44) requires that the direction of descent should not become nearly orthogonal to the gradient; Equation (5.45) indicates that the step length is of the same order of magnitude as the norm of the gradient with upper and lower bounds provided by the right and left inequalities, respectively. Within these restrictions there are a great many ways of choosing $s_n$ and $w_n$; we will discuss the most obvious, if perhaps not always the most efficient one.

Note that the functional $J$ decreases most rapidly in the direction of the negative gradient. Thus, we can generate $\{x_n\}$ by

$$x_{n+1} = x_n - t_n \nabla J(x_n). \tag{5.46}$$

This is known as the *method of a steepest descent*. It is of course of the form (5.42) with $w_n = \nabla J(x_n)/\|\nabla J(x_n)\|$ and $s_n = t_n \|\nabla J(x_n)\|$. Condition (5.44) is therefore satisfied with $k_1 = 1$ and (5.45) also simplifies and we immediately get Theorem 5.16.

**THEOREM 5.16.** If $t_n$ satisfies

$$\delta_1 \leq t_n \leq (2 - \delta_2)/K \tag{5.47}$$

for some positive $\delta_1, \delta_2 < 2$, then the method of steepest descent generates a critical sequence.

If the sequence $\{t_n\}$ is chosen within these limits then we obtain a critical sequence. A particularly simple way is to take the $t_n$ as constant, but this sometimes results in rather slow convergence (Exercise 6). Greater efficiency can often be obtained by minimizing $J$ in the direction of steepest descent by choosing $t_n = t_n^*$ such that

$$J(x_n - t_n^* \nabla J(x_n)) \leq J(x_n - t \nabla J(x_n)), \qquad \text{for all } t. \tag{5.48}$$

This requires a one-dimensional minimization at each step, but efficient procedures for this are known (e.g., various search techniques).

**THEOREM 5.17.** If $\|J''(x)\| \leq K$ for all $x \in X$, and $t_n^*$ is determined by (5.48), then the resulting sequence is critical.

**PROOF.** Assuming again that $\{x_n\}$ is not a critical sequence, there exists a subsequence $\{x_{n_j}\}$ such that

$$\| \nabla J(x_{n_j})\| \geq \mu > 0.$$

For any $t_{n_j}$ satisfying (5.47) we can show, following the steps of Theorem 5.15, that

$$J(x_{n_j}) - J\left(x_{n_j} - t_{n_j} \nabla J(x_{n_j})\right) \geq \frac{\mu^2 \delta_1 \delta_2}{2} \qquad (5.49)$$

and hence

$$J(x_{n_j}) - J\left(x_{n_j} - t_{n_j}^* \nabla J(x_{n_j})\right) \geq \frac{\mu^2 \delta_1 \delta_2}{2}.$$

Since by construction the sequence $J(x_n)$ is nonincreasing it follows that

$$J(x_{n_{j+1}}) - J(x_{n_j}) \geq \frac{\mu^2 \delta_1 \delta_2}{2} > 0,$$

leading to the contradiction $\lim J(x_{n_j}) \to -\infty$. Therefore the sequence must be critical. ∎

In conclusion, let us point out that the simple results presented here can be extended and generalized in many ways:

1. The restrictions on the existence and boundedness of the higher derivatives can be relaxed. Also, the required conditions need be satisfied only in some neighborhood of a critical point and not over the whole of $X$, as assumed here.

2. We have considered only convergence to a critical point. In practice one is of course more concerned with convergence to a minimum. As already mentioned, convexity arguments can often be used to determine when a critical sequence is also minimal. The speed of convergence is another question of practical importance.

3. The direction of steepest descent is the most appropriate choice if $J(x_n)$ and $\nabla J(x_n)$ are the only information available. During a sequence of steps, however, one gradually gains more knowledge of $J(x)$ and methods which take the "history" of the minimization

process into account are often more efficient. Typical of this class of techniques is the so-called *conjugate gradient method*.

For more details on all of these points the reader is referred to the cited references.

### Exercises 5.4

1. Give an example for which $J'(x^*)=0$, $J''(x^*)x^2 \geq 0$ for all $x$, but $x^*$ is not a local minimum.
2. Verify the claim made in the first part of the proof of Theorem 5.12.
3. Assume that $J$ is convex but not strictly so.
   (a) Give an example having a nonunique local minimum.
   (b) If $x^*$ and $x^{**}$ are both local minima, show that $J(x^*)=J(x^{**})$.
4. (a) Give an example of a critical sequence which is not minimal.
   (b) Give an example of a minimal sequence which does not converge.
5. Show that

$$\|J'(x_n)\| = \|\nabla J(x_n)\|$$

and

$$|J'(x_n)\nabla J(x_n)| = \|\nabla J(x_n)\|^2.$$

6. Apply the method of steepest descent with constant $t_n = 0.1$ to the (rather trivial) example,

$$J(x) = x^4, \quad x \in R^1.$$

Starting with $x_0 = 0.1$, show that the sequence is critical. Estimate the number of iterations required to get $x_n = 10^{-4}$.

7. Prove the inequality (5.49).
8. In Theorem 5.13 show that there exists some $z \neq x^*$ in $b(x^*, r) \cap \hat{l}(x^*, x^{**})$.

# PART III

# SPECIAL TOPICS

In this section we investigate how the general results developed so far can be applied and extended in certain specific settings. The areas selected, while not the only ones pursued by numerical analysts, are important and are of considerable current interest. Much of the work is still in the developmental stage with new results appearing almost daily. We hope that the interested reader will find these brief discussions an easy introduction to further reading and perhaps even research.

The topics to be discussed were chosen not only for their importance, but also because they show the different types of approaches taken by the theoretical numerical analyst to develop new results. In Chapter 6 we take a particular class of problems (integral and integro-differential equations) to show how the stability question can be handled in a very general way. This is one of the best examples of the success of function-analytic techniques in numerical analysis. In Chapter 7 we revisit an old technique, Galerkin's method. Revived under the name of Finite Element Method it has become an extremely powerful numerical method, particularly in the solution of partial differential equations. A significant portion of the new work in computational mathematics, both in its theoretical and practical aspects, is devoted to the study of this area. Chapter 8 demonstrates one of the favorite activities of mathematicians everywhere—the extension of a known body of results to new areas previously ruled out by restrictive assumptions. In particular, we show how the stability-convergence and the asymptotic expansion questions, treated for the linear case in Chapter 4, can be generalized for the nonlinear case. Finally, in Chapter 9 we consider a problem not covered by anything we have said so far since it violates the basic assumption of well-posedness. At first sight it may appear that ill-posed problems are of no practical interest, but this is not so. There has been much recent activity in this area.

As already mentioned, much of this work, especially that described in Chapters 8 and 9, is far from fully developed. We hope that this necessarily short introduction will stimulate the reader to pursue some of these questions on his own.

# 6

# THE APPROXIMATE SOLUTION OF LINEAR OPERATOR EQUATIONS OF THE SECOND KIND

We have seen in Chapter 4 that in the analysis of approximation methods for linear operator equations the question of stability plays a major role so that it is natural to look for some simple and easily verifiable criteria sufficient for stability. For the general case this search has unfortunately not been completely successful, although there exist a number of techniques, such as the maximum principle used in Example 4.6, which are very useful in analyzing specific cases. However, if we restrict ourselves to cases where the operators have a certain structure we can hope to obtain more far-reaching results. In this chapter we will look at one such class of operator equations, the so-called *equations of the second kind*, of which the Fredholm integral equation

$$x(t) + \int_0^1 k(t,s)x(s)\,ds = y(t), \qquad 0 \le t \le 1, \qquad (6.1)$$

is the simplest example.

A great deal of work has been done in connection with this equation; we will present several possible approaches in Sections 6.2–6.4. A recent extension to more general equations will be given in Section 6.5. Before proceeding with this we need to introduce some new ideas.

## 6.1 COMPACT OPERATORS AND EQUATIONS OF THE SECOND KIND

**DEFINITION 6.1.** Let $S$ be a subset of a normed linear space $X$. Then

(a)   $S$ is said to be *bounded* if $\|x\| \le M < \infty$ for all $x \in S$.

(b)　$S$ is said to be *closed* if $\{x_n\} \in S$ and $x_n \to x \in X$ implies that $x \in S$.

(c)　$S$ is said to be *compact* if every infinite sequence in $S$ contains a convergent subsequence with limit in $S$.

The definition of a compact set given here is in a form particularly convenient for further work; several alternative, but essentially equivalent definitions are used in the literature [see for example (Anselone, 1971, p. 4)].

The following two theorems relating the three concepts just introduced are well-known and the proofs can be found in most books on analysis. The reader may find it instructive to prove them independently.

**THEOREM 6.1.**　If $S$ is compact, then it is closed and bounded.

**THEOREM 6.2.**　If $S$ is a closed and bounded subset of a finite-dimensional space, then it is compact.

In an infinite-dimensional space, however, a closed and bounded set is not necessarily compact as can be seen from the following example.

**Example 6.1.**　Let $X$ be the space $C[0, 1]$ with the maximum norm. Then the set

$$S = \{x \mid x \in X, \|x\| \le 1\}$$

is bounded by definition. To show that it is also closed is a simple exercise in calculus which we leave to the reader. Consider now the sequence $x_n = \sin nx$. Then $x_n \in S$, but clearly there exists no convergent subsequence. Therefore $S$ is not compact.

**DEFINITION 6.2.**　A linear operator $K : X \to Y$ is said to be *compact* or *completely continuous* if every bounded sequence $x_n \in X$ contains a subsequence $x_n'$ such that $Kx_n'$ converges to an element in $Y$. An equivalent definition is to say that a compact operator maps every bounded subset $S \subset X$ into a set $KS \subset Y$ whose closure is compact.

**DEFINITION 6.3.**　A linear operator $L$ is said to be of *finite rank n* if

$$\dim(\mathscr{R}(L)) = n.$$

**THEOREM 6.3.**　If $K$ is a bounded linear operator of finite rank then it is compact.

PROOF. Let $\{x_n\}$ be a bounded sequence. Then $y_n = Kx_n$ is a bounded sequence in a finite-dimensional space. The closure of $\{y_n\}$ is therefore a closed and bounded set and by Theorem 6.2 it is compact. ∎

**THEOREM 6.4.** Let $\{K_n\}$ be a sequence of compact operators from a Banach space $X$ into itself such that

$$\lim_{n \to \infty} \|K - K_n\| = 0.$$

Then $K$ is compact.

PROOF. Let $\{x_n\}$ be a bounded sequence in $X$ with $\|x_n\| \leq B$. Since $K_1$ is compact there exists a subsequence of $\{x_n\}$, say $\{x_n^{(1)}\}$ such that $K_1 x_n^{(1)}$ converges to some element in $X$, say $y^{(1)}$. Now since $K_2$ is compact the sequence $\{x_n^{(1)}\}$ contains a subsequence $\{x_n^{(2)}\}$ such that $K_2 x_n^{(2)}$ converges to some $y^{(2)}$ in $X$. By repeating this argument we see that for any $m$ there exists a subsequence $\{x_n^{(m)}\}$ such that $K_m x_n^{(m)}$ converges to $y^{(m)} \in X$. From these we now pick a sequence of distinct elements $z_m$ such that $z_m \in \{x_n^{(m)}\}$ and $\|K_m z_m - y^{(m)}\| \leq 1/m$. Then

$$\|Kz_m - Kz_{m+p}\|$$
$$\leq \|Kz_m - K_m z_m\| + \|K_m z_m - K_m z_{m+p}\| + \|K_m z_{m+p} - Kz_{m+p}\|.$$

Since $z_{m+p}$ is in $\{x_n^{(m)}\}$ we have

$$\lim_{p \to \infty} K_m z_{m+p} = y^{(m)}.$$

Thus

$$\lim_{p \to \infty} \|Kz_m - Kz_{m+p}\| \leq 2\|K - K_m\|B + \|K_m z_m - y^{(m)}\|$$

and

$$\lim_{m \to \infty} \lim_{p \to \infty} \|Kz_m - Kz_{m+p}\| = 0.$$

Therefore $\{Kz_m\}$ is a Cauchy sequence, and because of the completeness of $X$ the sequence converges to some element in $X$. Thus $K$ is compact. ∎

The reason for presenting the last two theorems is that we wish to consider integral operators of the form

$$(Kx)(s) = \int_0^1 k(s,t)x(t)\,dt. \tag{6.2}$$

For certain types of kernels this defines a compact operator.

**THEOREM 6.5.**   Let the kernel $k(s,t)$ in (6.2) be a continuous function of $s$ and $t$ in $[0,1] \times [0,1]$. If $K$ is considered as an operator $X \to X$, where $X$ is $C[0,1]$ with maximum norm, then $K$ is compact.

PROOF.   According to the Weierstrass theorem there exist functions

$$k_n(s,t) = \sum_{i=0}^{n_1} \sum_{j=0}^{n_2} c_{ij} s^i t^j,$$

such that $|k(s,t) - k_n(s,t)|$ can be made as small as desired by choosing $n_1$ and $n_2$ sufficiently large. The operators $K_n$ defined by

$$(K_n x)(s) = \int_0^1 k_n(s,t) x(t) \, dt \tag{6.3}$$

are of finite rank $n_1 + 1$. Furthermore, from Example 1.11

$$\|K - K_n\| = \max_{0 \le s \le 1} \int_0^1 |k(s,t) - k_n(s,t)| \, dt$$

so that

$$\lim_{n \to \infty} \|K - K_n\| = 0.$$

Applying Theorems 6.3 and 6.4 then completes the proof.   ■

One can also prove compactness for this type of integral operator by applying the well-known *Arzela–Ascoli theorem.*

**THEOREM 6.6.**   Let $B$ be a subset of $C[D]$, $D \in R^k$, with maximum norm. If the set of functions $f \in B$ is uniformly bounded and equicontinuous then $B$ has compact closure.

PROOF.   See (Graves, 1956, p. 122).   ■

We leave it as an exercise to show how Theorem 6.5 follows from the Arzela–Ascoli theorem. The converse of this result is Theorem 6.7.

**THEOREM 6.7.**   Let $B$ be a compact subset of $C[D]$ as defined above. Then the set of functions $f \in B$ is bounded and equicontinuous.

PROOF.   See (Graves, 1956, p. 122).   ■

**Example 6.2.**   While the identity operator on a finite-dimensional space is compact, this is no longer true in infinite-dimensional spaces. In some sense compact operators are smoothing, in that they map functions possessing certain smoothness properties into others with more smoothness (Exercise 6).

Closely connected with the notion of a compact operator is the following idea.

**DEFINITION 6.4.**  Let $\{X_n\}$ be a sequence of normed linear spaces and let $\{K_n\}$ be a sequence of operators $X_n \to X$. Then the set $\{K_n\}$ is said to be *collectively compact* if for every bounded sequence $x_n \in X_n$ the sequence $\{K_n x_n\}$ contains a subsequence converging to some element in $X$.

This notion was first used in the work of Anselone (Anselone, 1971). The definition given here is a slight generalization of that of Anselone to cover the case when the domains of the $K_n$ are not the same space. This extension will be needed in Section 6.5.

The particular form (6.2) for the integral operator was chosen for simplicity; it should be obvious that an integral operator over a general region in $R^k$ is compact if the kernel is continuous. Also, the restriction of continuity can be removed to a certain extent. A rather simple special case of this is given in Exercise 4, but we will not pursue this further. For the rest of this chapter we will assume that the underlying space is the Banach space $C[0, 1]$ with maximum norm, and that the kernels are continuous on $[0, 1] \times [0, 1]$.

The Fredholm integral equation (6.1) can then be written in operator notation as

$$(\lambda I + K)x = y, \tag{6.4}$$

where $K$ is a compact operator $X \to X$. We will always assume that $\lambda$ is not zero and that $(\lambda I + K)^{-1}$ exists and is bounded. We will consider approximate methods for this type of equation in some detail in Sections 6.2–6.4. In Section 6.5 we will consider a generalization of the form

$$(\lambda A + K)x = y, \tag{6.5}$$

where it will be assumed that

(a)  $A$ is a linear, but not necessarily bounded operator whose domain and range are in a Banach space $X$.
(b)  $A^{-1}$ exists and is bounded in $X$.
(c)  $K$ is a compact operator $X \to X$.
(d)  $\lambda$ is such that $(\lambda A + K)$ has a bounded inverse on $X$.

An equation of the form (6.5) will be called a general *equation of the second kind*. This equation is of course equivalent to one of the simpler form (6.4) since we can rewrite it as

$$(\lambda I + A^{-1}K)x = A^{-1}y. \tag{6.6}$$

However, in many cases of practical interest $A^{-1}$ is not known, so that it is necessary to look at (6.5) directly.

### Exercises 6.1

1. Show that every compact operator is bounded.
2. If $K_1$ is a bounded and $K_2$ is a compact operator, show that $K_1 K_2$ and $K_2 K_1$ are compact.
3. Prove Theorems 6.1 and 6.2.
4. Assume that the kernel $k(s,t)$ in (6.2) is continuous in $s$, but has a simple jump discontinuity along the line $t = t_1$. Show that $K$ is compact.
5. Prove that the closed unit ball in an infinite-dimensional Hilbert space cannot be compact.
6. Let $k(s,t)$ be bounded and continuous on $[0,1] \times [0,1]$. Show that the integral operator defined by (6.2) maps $L_2[0,1]$ into $C[0,1]$.
7. Show that the integro-differential equation

$$\lambda x''(s) + \int_0^1 k(s,t)x(t)\,dt = y(s)$$
$$x(0) = \alpha, \quad x(1) = \beta,$$

   is an equation of the second kind.
8. Show how Theorem 6.5 follows from the Arzela–Ascoli theorem.
9. Prove that a compact operator on a Hilbert space cannot have a bounded inverse unless it is of finite rank.

### 6.2  THE METHOD OF DEGENERATE KERNELS FOR FREDHOLM EQUATIONS

For Fredholm integral equations of the second kind Theorem 6.5 suggests a simple approximation method. In general, we approximate the kernel $k(s,t)$ by a finite-rank kernel of the form

$$k_n(s,t) = \sum_{i=0}^{n_1} \sum_{j=0}^{n_2} c_{ij}\varphi_i(s)\psi_j(t), \tag{6.7}$$

where $\{\varphi_i\}$ and $\{\psi_i\}$ are linearly independent sets of expansion functions. We then look for an approximate solution $x_n(s)$ satisfying

$$\lambda x_n + K_n x_n = y. \tag{6.8}$$

Written out explicitly this equation is

$$\lambda x_n(s) = y(s) - \int_0^1 \sum_{i=0}^{n_1} \sum_{j=0}^{n_2} c_{ij} \varphi_i(s) \psi_j(t) x_n(t) \, dt, \tag{6.9}$$

so that

$$\lambda x_n(s) = y(s) - \sum_{i=0}^{n_1} \gamma_i \varphi_i(s), \tag{6.10}$$

with

$$\gamma_i = \sum_{j=0}^{n_2} c_{ij} \int_0^1 \psi_j(t) x_n(t) \, dt. \tag{6.11}$$

If we substitute the expansion (6.10) into (6.11) we obtain a linear system for the $\gamma_i$,

$$\gamma_i = \frac{1}{\lambda} \sum_{j=0}^{n_2} c_{ij} \int_0^1 \psi_j(t) \left\{ y(t) - \sum_{k=0}^{n_1} \gamma_k \varphi_k(t) \right\} dt, \qquad i = 0, 1, \ldots, n_1.$$

$$\tag{6.12}$$

If the integrals in the above expression can be evaluated, then the approximate solution can be obtained by solving a linear system.

To show that an $x_n$ solving (6.8) exists and to obtain a convergence theorem and error bounds we must investigate the solvability of (6.8).

**THEOREM 6.8.** If

$$\|K - K_n\| \le 1 / \|(\lambda I + K)^{-1}\|, \tag{6.13}$$

then $(\lambda I + K_n)^{-1}$ exists and

$$\|(\lambda I + K_n)^{-1}\| \le \frac{\|(\lambda I + K)^{-1}\|}{1 - \|(\lambda I + K)^{-1}\| \, \|K - K_n\|}. \tag{6.14}$$

PROOF. This is just a special case of Theorem 4.5 with $L = \lambda I + K$ and $\Delta L = K_n - K$. Therefore, if $\lim \| K - K_n \| = 0$, then for sufficiently large $n$ (6.8) has a unique solution and the approximating method is stable. ■

**THEOREM 6.9.** If $x$ and $x_n$ are the solutions of (6.4) and (6.8), respectively, and if (6.13) is satisfied, then

$$\| x - x_n \| \le \frac{\| (\lambda I + K)^{-1} \| \, \| K - K_n \| \, \| x \|}{1 - \| (\lambda I + K)^{-1} \| \, \| K - K_n \|}. \tag{6.15}$$

PROOF. Subtracting (6.8) from (6.4) we have

$$(\lambda I + K_n)(x - x_n) = K_n x - Kx,$$

$$\| x - x_n \| \le \| (\lambda I + K_n)^{-1} \| \, \| K - K_n \| \, \| x \|.$$

Using (6.14) the bound follows. ■

**THEOREM 6.10.** If $\| K - K_n \| \le 1 / \| (\lambda I + K_n)^{-1} \|$, then

$$\| x - x_n \| \le \frac{\| (\lambda I + K_n)^{-1} \| \, \| K - K_n \| \, \| x_n \|}{1 - \| (\lambda I + K_n)^{-1} \| \, \| K - K_n \|}. \tag{6.16}$$

PROOF. Reversing the roles of $K$ and $K_n$ in Theorem 6.8 we get

$$\| (\lambda I + K)^{-1} \| \le \frac{\| (\lambda I + K_n)^{-1} \|}{1 - \| (\lambda I + K_n)^{-1} \| \, \| K - K_n \|}.$$

From (6.8) and (6.4)

$$(\lambda I + K)(x - x_n) = (K - K_n)x_n$$

and (6.16) follows. ■

The error bounds (6.15) and (6.16), in spite of their similarity, differ in some important aspects. If $\lim \| K - K_n \| = 0$, then (6.15) guarantees that $x_n$ converges to $x$, but an explicit error bound is not immediately available from it since this requires bounds on $\| x \|$ and $\| (\lambda I + K)^{-1} \|$. Often this information, which must come from other considerations, is not available. Therefore (6.15) is primarily useful for proving convergence and finding the order of the method. The bound (6.16), on the other hand, cannot be used directly to establish convergence since we do not have bounds on $\|x_n\|$ and $\|(\lambda I + K_n)^{-1}\|$. Nevertheless, (6.16) is useful since for a given $n$ it is computable. The quantities $\|x_n\|$ and $\|(\lambda I + K_n)^{-1}\|$ can be obtained as part of the computation and $\|K - K_n\|$ can usually be found using results from approximation theory. Thus, for a given $n$, $\|x - x_n\|$ can be bounded. The inequality (6.16) is what we have previously called an *a posteriori* error bound.

The form of the degenerate kernel approximation (6.7) is arbitrary, but for practical reasons one normally uses simple expansion functions such as polynomials, piecewise polynomials, or splines. Once a form for $k_n$ has been chosen one must then evaluate the required integrals. Finally, the linear system (6.12) must be solved. The whole procedure tends to be somewhat cumbersome, at least more so than some of the alternatives. The main advantage of the degenerate kernel method seems to lie in the fact that it is intuitively simple and easily analyzed.

The simplicity of the error analysis for the degenerate kernel method arises to a large degree from its rather special properties. The approximate operators $K_n$ and the solutions $x_n$ are in the same space as the original problem; moreover, the operators $K_n$ can be considered as derived from $K$ by a bounded perturbation. It should be realized that this is a very special situation and that approximation methods of this type are rare. Certainly the usual methods for differential equations cannot be considered this way. Even in the solution of integral equations most methods do not have this property.

## Exercises 6.2

1. Show that if $\|K\| < |\lambda|$, then it is possible to bound $\|(\lambda I + K)^{-1}\|$ and $\|x\|$, so that we can use (6.15) to bound the error.
2. Investigate the problems encountered in the use of the degenerate kernel when $k(s,t)$ is approximated by polynomials or piecewise polynomials. What can be said about the order of convergence of such approaches? What are the practical difficulties encountered?

## 6.3  THE NYSTRÖM METHOD

One technique which immediately suggests itself is to use a numerical integration rule to replace the integral in (6.1). Thus, for a given $n$, we introduce the points $0 \le t_{n1} < t_{n2} < \cdots \le t_{nn} \le 1$ and replace the integral with a quadrature on these points. This leads to the equation

$$\lambda x_n(s) + \sum_{j=1}^{n} w_{nj} k(s, t_{nj}) x_n(t_{nj}) = y(s), \qquad (6.17)$$

where the $w_{nj}$ are the appropriate quadrature weights. It may not be immediately clear that such a system can have a solution, but if we satisfy (6.17) at the points $s = t_{ni}$, then

$$\lambda x_n(t_{ni}) + \sum_{j=1}^{n} w_{nj}k(t_{ni}, t_{nj})x_n(t_{nj}) = y(t_{ni}), \qquad i = 1, 2, \ldots, n. \quad (6.18)$$

This is a system of $n$ linear equations in $n$ unknowns and has therefore a unique solution except for special values of $\lambda$. We also note that (6.17) and (6.18) are equivalent: any solution of (6.17) also satisfies (6.18) and any solution of (6.18) defines a solution of (6.17) by simply substituting $x_n(t_{nj})$ in the summed term on the left-hand side. This observation was first made by Nyström (1930) in his study of methods of this type. Equation (6.17) is sometimes called the natural interpolation formula, since it defines $x_n(t)$ for all $t$ given $\{x_n(t_{ni})\}$.

The actual numerical solution of the Fredholm integral equation is carried out using (6.18), but the analysis in this section will be done considering (6.17). Therefore the Nyström observation is essential for what is to follow.

In operator notation we write (6.17) as

$$\lambda x_n + K_n x_n = y, \qquad (6.19)$$

where $x_n \in X$, $K_n : X \to X$, while (6.18) becomes

$$\lambda \mathbf{x}_n + \mathbf{K}_n \mathbf{x}_n = \mathbf{y}, \qquad (6.20)$$

with $\mathbf{x}_n, \mathbf{y} \in R^n, \mathbf{K}_n : R^n \to R^n$.

A great deal of work has been done on Nyström-type methods in the last few years; the major results are described in (Anselone, 1971). What makes the study of the Nyström method more difficult than the degenerate kernel method is the fact that we can no longer claim that $\lim \|K - K_n\| = 0$, as was pointed out in Section 3.3. Therefore new techniques are required to treat this case. The Anselone theory, which provides one possible approach, will now be briefly outlined.

**DEFINITION 6.5.** A subset $S$ of a normed space $X$ is said to be *totally bounded* if for every $\epsilon > 0$ it contains an $\epsilon$-net, that is, for every $\epsilon > 0$ there exists a finite subset $E$ of $S$ such that for every $x \in S$ there is an $e \in E$ for which

$$\|x - e\| \le \epsilon.$$

**THEOREM 6.11.** A subset $S$ of a Banach space is compact if and only if it is totally bounded and closed.

PROOF. We leave this as an exercise. For typical arguments used in establishing such results see (Graves, 1956, p. 358). ∎

**THEOREM 6.12.** Let $K_n : X \to X$ be a uniformly bounded sequence of linear operators converging pointwise to a bounded linear operator $K$, by which we mean that for every $x \in X$

$$K_n x \to Kx.$$

Then the convergence is uniform on compact subsets of $X$.

PROOF. Let $S$ be a compact subset of $X$. Then by Theorem 6.11, for every $\epsilon > 0$ there exists an $\epsilon$-net for $S$ consisting of elements $e_1, e_2, \ldots, e_m$. Now for $x \in S$

$$\|K_n x - Kx\| \leq \|K_n e_i - Ke_i\| + \|K_n x - K_n e_i\| + \|Ke_i - Kx\|, \qquad \text{for all } i.$$

Since there are only a finite number of the $e_i$, we can make $n$ large enough so that the first term on the right-hand side is smaller than $\epsilon$. Then

$$\|K_n x - Kx\| \leq \epsilon + \|K_n\|\epsilon + \|K\|\epsilon,$$

so that

$$\sup_{x \in S} \|K_n x - Kx\| \leq c\epsilon,$$

and the convergence is uniform in $S$. ∎

**THEOREM 6.13.** Let $L$ be a compact operator and let $K_n$ be a uniformly bounded sequence of linear operators converging pointwise to some operator $K$. Then

$$\lim_{n \to \infty} \|(K - K_n)L\| = 0. \tag{6.21}$$

PROOF.

$$\|(K - K_n)L\| = \sup_{\|x\| = 1} \|(K - K_n)Lx\|.$$

Since $L$ is compact the closure of the set $S = \{y \,|\, y = Lx, \|x\| = 1\}$ is compact. Hence by Theorem 6.12, convergence is uniform on $S$ and

$$\lim_{n\to\infty} \sup_{\|x\|=1} \|(K-K_n)Lx\| = 0,$$

proving (6.21).    ∎

**THEOREM 6.14.** Let $\{K_n\}$ be a uniformly bounded sequence of linear operators converging pointwise to $K$, and let $\{L_n\}$ be a sequence of collectively compact operators. Then

$$\lim_{n\to\infty} \|(K-K_n)L_n\| = 0. \tag{6.22}$$

**PROOF.** Since the sequence $\{L_n\}$ is collectively compact and the set $S=\{y \mid y=L_n x, \|x\|=1\}$ is compact, the proof is essentially identical to that of the previous theorem.    ∎

**THEOREM 6.15.** Let $L$ and $M$ be bounded linear operators and assume that $(\lambda I - L)^{-1}$ exists and is bounded. Furthermore, assume that if $\mu \notin \sigma(M)$, then $(\mu I - M)^{-1}$ exists and is bounded. Then if

$$\|(M-L)M\| < |\lambda| / \|(\lambda I - L)^{-1}\|,$$

$(\lambda I - M)^{-1}$ exists and

$$\|(\lambda I - M)^{-1}\| \le \frac{\|I+(\lambda I - L)^{-1}M\|}{|\lambda| - \|(\lambda I - L)^{-1}\| \, \|(M-L)M\|}. \tag{6.23}$$

**PROOF.** Note that

$$\lambda(\lambda I - L)^{-1} = (\lambda I - L)^{-1}(\lambda I - L + L) = I + (\lambda I - L)^{-1}L.$$

Also

$$\begin{aligned}
\lambda I - (\lambda I - L)^{-1}(M-L)M &= \lambda I - (\lambda I - L)^{-1}M^2 + (\lambda I - L)^{-1}LM \\
&= [I+(\lambda I - L)^{-1}L]M - (\lambda I - L)^{-1}M^2 + \lambda I - M \\
&= \lambda(\lambda I - L)^{-1}M - (\lambda I - L)^{-1}M^2 + \lambda I - M \\
&= (\lambda I - L)^{-1}M(\lambda I - M) + \lambda I - M \\
&= [I+(\lambda I - L)^{-1}M](\lambda I - M).
\end{aligned}$$

Therefore the operator $[I + (\lambda I - L)^{-1} M](\lambda I - M)$ is of the form $\lambda(I - A)$ with $\|A\| < 1$. It therefore has a bounded inverse. This implies that $\lambda \notin \sigma(M)$ so that by assumption $(\lambda I - M)^{-1}$ exists and is bounded. Inequality (6.23) follows by writing

$$(\lambda I - M)^{-1} = \frac{1}{\lambda}\left[ I - \frac{1}{\lambda}(\lambda I - L)^{-1}(M - L)M \right]^{-1}\left[ I + (\lambda I - L)^{-1} M \right]$$

and applying (4.12) to the first part of the right-hand side. ∎

We can now prove convergence for Nyström's method.

**THEOREM 6.16.** Let the operators $K_n$ in (6.19) be uniformly bounded and collectively compact. Let $K$ and $x$ be as defined in (6.4) and assume that $K_n$ converges to $K$ pointwise. Then for sufficiently large $n$, (6.19) has a unique solution with $x_n \to x$. Furthermore

$$\|x - x_n\| \leq \frac{1 + \|(\lambda I + K)^{-1}\| \|K_n\|}{|\lambda| - \|(\lambda I + K)^{-1}\| \|(K - K_n)K_n\|} \|(K - K_n)x\|. \quad (6.24)$$

PROOF. From Theorem 6.14 with $L_n = K_n$ we can always choose $n$ large enough so that

$$\|(K - K_n)K_n\| \leq |\lambda| / \|(\lambda I + K)^{-1}\|.$$

Use now Theorem 6.15 with $L = -K$ and $M = -K_n$. Since $K_n$ is compact the conditions imposed on it in that theorem are satisfied. Then, for sufficiently large $n$, $(\lambda I + K_n)^{-1}$ exists and

$$\|(\lambda I + K_n)^{-1}\| \leq \frac{1 + \|(\lambda I + K)^{-1}\| \|K_n\|}{|\lambda| - \|(\lambda I + K)^{-1}\| \|(K - K_n)K_n\|}. \quad (6.25)$$

The Nyström method is therefore stable and (6.24) follows by subtracting (6.19) from (6.4). ∎

To apply the theorem to specific cases we must verify the assumptions that the $K_n$ are uniformly bounded and collectively compact. If we impose the plausible restriction

$$\sup_n \sum_{i=1}^{n} |w_{ni}| \leq W < \infty \quad (6.26)$$

then since we also assumed that $k(s, t)$ is bounded, the $K_n$ are uniformly bounded. Now let $\{x_n\}$ be a sequence of elements with $\|x_n\| \le M$. Then

$$(K_n x_n)(s_1) - (K_n x_n)(s_2) = \sum_{j=1}^{n} w_{nj}\big(k(s_1, t_{nj}) - k(s_2, t_{nj})\big)x_n(t_{nj}),$$

$$|K_n x_n(s_1) - K_n x_n(s_2)| \le WM \max_{0 \le t \le 1} |k(s_1, t) - k(s_2, t)|.$$

Thus the set $\{K_n x_n | \|x_n\| \le M\}$ is uniformly bounded and equicontinuous, so by the Arzela–Ascoli theorem it is compact. By definition then the sequence $\{K_n\}$ is collectively compact.

Continuity of $k(s, t)$ together with (6.26) then assures the convergence (and stability) of the Nyström method. Continuity is, however, not necessary and the method converges under less restrictive conditions. This aspect is treated in detail in Anselone's book.

Again, as in Section 6.2, we can obtain *a posteriori* error bounds. By reversing the roles of $K$ and $K_n$ in Theorem 6.16 we see that

$$\|x - x_n\| \le \frac{1 + \|(\lambda I + K_n)^{-1}\| \, \|K\|}{|\lambda| - \|(\lambda I + K_n)^{-1}\| \, \|(K - K_n)K\|} \|(K - K_n)x_n\|, \quad (6.27)$$

provided that $\|(K - K_n)K\| \le |\lambda| / \|(\lambda I + K_n)^{-1}\|$. This can be used to bound the error once an approximate solution has been obtained.

### Exercises 6.3

1. Show how, in general, it is possible to bound $\|(K - K_n)K\|$ and $\|(K - K_n)K_n\|$.
2. Prove (6.27).
3. Consider the integral equation

$$x(t) + \int_0^1 e^{st}x(s)\,ds = 1, \qquad 0 \le t \le 1.$$

   If we use the Nyström method based on the composite Simpson's rule
   (a)  Compute bounds for $\|K_n\|$, $\|(K - K_n)K\|$ and $\|(K - K_n)K_n\|$.
   (b)  Establish the order of convergence of the method.
4. If $k(s, t)$ is continuous everywhere in $[0, 1] \times [0, 1]$ except along the line $s = t$ where it has a simple jump discontinuity, show that
   (a)  $K$ is compact.
   (b)  Propose a method of solution analogous to the Nyström method and show that the $K_n$ are collectively compact.

(c)  Prove that the method converges and establish the order of convergence.

5.  Prove Theorem 6.11.

## 6.4  ANALYSIS OF METHODS FOR INTEGRAL EQUATIONS VIA RESTRICTIONS AND PROLONGATIONS

The analysis in the previous section was made possible by the existence of the natural interpolation formula, that is, by the equivalence of (6.19) and (6.20). While the numerical solution is of course computed by solving (6.20), the error analysis uses (6.19) and thus allows us to ignore the fact that $x$ and $x_n$ are not in the same space. In spite of the theoretical elegance of this approach, it has the drawback that it is restricted as to the type of the equation and the form of the approximation. To generalize somewhat let us consider an approximating equation of the form

$$\lambda x_n + K_n x_n = y_n, \tag{6.28}$$

without specifying the exact form of $K_n$. All we shall assume is that $x_n$ and $y_n$ are elements of a finite-dimensional space $x_n$, and that $K_n$ is a linear operator $X_n \rightarrow X_n$. As we have already pointed out in Chapter 4, we must establish a connection between the spaces $X$ and $X_n$; to do this we use the previously introduced restrictions and prolongations with the assumption that they satisfy the conditions stated in Chapter 4. The basic idea outlined here is due to Noble (1973).

Before presenting the main theorem we need to generalize Theorem 4.4 dealing with the existence of certain inverses.

**THEOREM 6.17.**  A linear operator $L:X \rightarrow X$ has a left inverse $L_L^{-1}$ if there exists an operator $K \in L[X,X]$ such that

$$\|I - KL\| < 1.$$

If this is satisfied then

$$\|L_L^{-1}\| \leq \frac{\|K\|}{1 - \|I - KL\|} . \tag{6.29}$$

PROOF.  Let $M = I - KL$. Then $\|M\| < 1$, so that by Theorem 4.5 $(I - M)$ has a bounded inverse. Now $KL = I - M$ and

$$(I - M)^{-1} KL = I.$$

Hence

$$L_L^{-1} = (I-M)^{-1}K$$

and (6.29) follows from Theorem 4.5. ∎

A convergence theorem for methods of the type (6.28) can now be obtained.

**THEOREM 6.18.** Let $K$ be a compact operator, $\{K_n\}$ a sequence of linear operators $X_n \to X_n$, and let $r_n$ and $p_n$ satisfy the usual conditions. If

$$\lim_{n\to\infty} \|K_n - r_n K p_n\| = 0, \tag{6.30}$$

then the approximation method (6.28) is stable and, for sufficiently large $n$,

$$\|(\lambda I_n + K_n)^{-1}\| \le \left(\|(\lambda I + K)^{-1}\|\|p_n\|\|r_n\|\right)/(1-\alpha_n), \tag{6.31}$$

where

$$\alpha_n = \|r_n\|\|(\lambda I + K)^{-1}\|\|p_n\|\{\|K_n - r_n K p_n\| + \|(I - p_n r_n)K\|\}.$$

PROOF. Consider

$$I_n - r_n(\lambda I + K)^{-1}p_n(\lambda I_n + K_n)$$

$$= r_n(\lambda I + K)^{-1}[(\lambda I + K)p_n - p_n(\lambda I_n + K_n)]$$

$$= r_n(\lambda I + K)^{-1}[Kp_n - p_n K_n]$$

$$= r_n(\lambda I + K)^{-1}[(I - p_n r_n)Kp_n - p_n(K_n - r_n K p_n)].$$

We want to show that the right-hand side can be made arbitrarily small. Now

$$\|(I - p_n r_n)K\| = \sup_{\|x\|=1} \|(I - p_n r_n)Kx\|$$

$$= \sup_{z \in B} \|(I - p_n r_n)z\|,$$

where $B = \{z \,|\, z = Kx, \|x\| = 1\}$. By (4.39) it follows that for all $z \in X$

$$\lim_{n\to\infty} \|(I - p_n r_n)z\| = 0,$$

that is, $I - p_n r_n$ converges to zero pointwise. By Theorem 6.12 this convergence is uniform on compact subsets; therefore since $B$ is compact we have

$$\lim_{n \to \infty} \|(I - p_n r_n)K\| = 0.$$

From this and (6.30) it follows that for sufficiently large $n$

$$\|I_n - r_n(\lambda I + K)^{-1} p_n(\lambda I_n + K_n)\| \leq \alpha_n < 1.$$

Applying Theorem 6.17 then shows that $(\lambda I_n + K_n)$ has a left inverse. Since this is an operator from the finite-dimensional space $X_n$ into itself it can be represented by a square matrix and the existence of a left inverse implies the existence of an inverse. Thus the $(\lambda I_n + K_n)^{-1}$ exist and are uniformly bounded so that the method is stable. ∎

This result is quite general since no special requirements, other than those on $p_n$ and $r_n$, have been made. For example, the method of degenerate kernels is within the scope of this theorem. Here $X_n = X$, so that we can take $r_n$ and $p_n$ as the identity operator. Condition (6.30) is of course satisfied for an appropriate choice of expansion functions.

For the Nyström method we can define $r_n : X \to R^n$ by

$$r_n x = \begin{bmatrix} x(t_{n1}) \\ x(t_{n2}) \\ \vdots \\ x(t_{nn}) \end{bmatrix} \tag{6.32}$$

which is an obvious choice, implying that $y_n = r_n y$, so that stability and consistency guarantee convergence. The choice of $p_n$, on the other hand, is not imposed on us by the method. Theorem 6.18 must be interpreted as saying that if there exists any appropriate $p_n$ such that (6.30) is satisfied, then the method is stable. While the natural interpolation (6.17) defines a prolongation, this operator does not satisfy all the restrictions we have imposed (Exercise 1). To get around this we must define another $p_n$, one that satisfies (6.30) as well as the conditions stated in Chapter 4. This can be done in many ways; a simple example will suffice to demonstrate the general principle.

**Example 6.3.** If in the Nyström method we use equidistant points $t_{ni} = (i - 1)h$, $h = 1/(n - 1)$ and weights $w_{n1} = w_{nn} = h/2$, $w_{ni} = h, i = 2, 3, \ldots, n - 1$, then we are effectively using the composite trapezoidal method to replace the integral in (6.1). As $p_n$ we take the operators of linear interpolation, so that with each $x_n$ is associated a piecewise linear function taking on the value $x_{ni}$ at $t_{ni}$. To get a specific representation for $p_n$ we introduce the "hat" functions

$$\varphi_{ni}(t) = (t - t_{n,i-1})/h, \qquad 0 \le t_{n,i-1} \le t \le t_{ni},$$
$$= (t_{n,i+1} - t)/h, \qquad t_{ni} \le t \le t_{n,i+1} \le 1,$$
$$= 0, \qquad \text{otherwise.}$$

Then

$$(p_n x_n)(t) = \sum_{i=1}^{n} \varphi_{ni}(t) x_{ni}.$$

Using the maximum norm we have

$$\|r_n K p_n - K_n\| = \max_i \left| \sum_{j=1}^{n} w_{nj} k(t_{ni}, t_{nj}) - \int_0^1 k(t_{ni}, t) \sum_{j=1}^{n} \varphi_{nj}(t)\, dt \right|.$$

If $k(t,s)$ is twice continuously differentiable in its second argument, then a few manipulations (Exercise 3) yield

$$\|r_n K p_n - K\| \le \frac{h^2}{6} \max_i \left\{ \left| \frac{\partial k}{\partial t}(t_{ni}, 0) \right| + \left| \frac{\partial k}{\partial t}(t_{ni}, 1) \right| + \max_t \left| \frac{\partial^2 k}{\partial t^2}(t_{ni}, t) \right| \right\}$$

$$(6.33)$$

Therefore (6.30) is satisfied.

We will not pursue this question of checking (6.30) in other cases since in the next section we obtain a more general and more easily verifiable criterion.

### Exercises 6.4

1. Show that if $r_n$ is defined by (6.32) and $p_n$ is given by (6.17), then not all the conditions (4.36)–(4.40) are satisfied.
2. Show that if $r_n$ is defined by (6.32) and $p_n$ is as defined in Example 6.3, then all the required conditions are met.
3. For the approach outlined in Example 6.3
   (a) Show that $\int_0^1 \varphi_{ni}(s)\, ds = w_{ni}$
   (b) Show that

$$\|r_n K p_n - K_n\| \le \max_{1 \le i \le n} \sum_{j=1}^{n} \left| \int_0^1 \left( k(t_{ni}, t) - k(t_{ni}, t_{nj}) \right) \varphi_{nj}(t)\, dt \right|.$$

(c)  By expanding $k(t_{ni}, t)$ in a Taylor series about $t_{nj}$ deduce (6.33).

(d)  If $k(t, s)$ is twice differentiable in the first argument, show that

$$\|(I - p_n r_n)K\| \leq \frac{h^2}{8} \max_{s,t} \left| \frac{\partial^2}{\partial s^2} k(s, t) \right|.$$

4.  Show that if a square matrix $A$ has a left inverse, then it is nonsingular and $A^{-1} = A_L^{-1}$.

## 6.5  GENERAL OPERATOR EQUATIONS OF THE SECOND KIND

We now return to the general problem (6.5). The discretized version of this equation can be written as

$$(\lambda A_n + K_n)x_n = y_n, \tag{6.34}$$

with $x_n, y_n \in X_n$, where $X_n$ is a finite-dimensional space. The methods of analyzing integral equations presented in the previous sections do not generalize immediately to this case since, in one way or another, they depend on the fact that the equation was of the form of an identity plus a compact operator. To remedy this situation we will combine certain of the ideas of Sections 6.3 and 6.4 to produce a more generally applicable result.

Still, in order to make any progress, we must make some assumptions. We cannot prove stability without further restrictions since this would be tantamount to showing that all consistent discretizations of well-posed problems are stable, which is of course not true. We therefore assume that $A$ is relatively simple, so that $\{A_n\}$ can be shown to be stable. It is then possible to prove that, under fairly general conditions, $\{\lambda A_n + K_n\}$ is stable. As we indicate in Example 6.4 this allows one to treat a wide variety of integro-differential equations.

We start by deriving a few useful results.

**THEOREM 6.19.** If $\{L_n, r_n, p_n\}$ is a stable and consistent approximation of $L$, then $\{L_n^{-1}, r_n, p_n\}$ is consistent with $L^{-1}$ (assuming that $L$ is invertible).

**PROOF.**  We want to show that for any $y \in \mathfrak{D}(L^{-1})$

$$\lim_{n \to \infty} \|r_n L^{-1} y - L_n^{-1} r_n y\| = 0.$$

Since $L^{-1}$ exists and $\{L_n\}$ is stable, it follows that for any $y \in \mathfrak{D}(L^{-1})$

there exists an $x$ and a sequence $\{x_n\}$ such that, for sufficiently large $n$,

$$Lx = y$$

and

$$L_n x_n = r_n y.$$

Since $\{L_n, r_n, p_n\}$ is a stable and consistent approximation of $L$, we have by Theorem 4.13 that

$$\lim_{n \to \infty} \|r_n x - x_n\| = 0.$$

Now

$$r_n x = r_n L^{-1} y$$

and

$$x_n = L_n^{-1} r_n y,$$

so that

$$\|r_n L^{-1} y - L_n^{-1} r_n y\| \to 0. \qquad \blacksquare$$

**THEOREM 6.20.** Let $\{L_n, r_n, p_n\}$ be consistent with the bounded linear operator $L$, such that $\{L_n\}$ is uniformly bounded. Let $\{M_n, r_n, p_n\}$ be consistent with the bounded linear operator $M$. Then $\{L_n M_n, r_n, p_n\}$ is consistent with $LM$, provided that all operator products are defined.

PROOF.   This is a simple exercise which will be left to the reader.   $\blacksquare$

**THEOREM 6.21.** Let $\{L_n, r_n, p_n\}$ be consistent with the bounded linear operator $L$, such that $\{L_n\}$ is uniformly bounded. If $\{x_n\}$ is a sequence such that

$$p_n x_n \to x \in X,$$

then

$$\lim_{n \to \infty} \|(\lambda I_n + L_n) x_n\| = \|(\lambda I + L) x\|. \qquad (6.35)$$

PROOF. Since by assumption $x_n$ converges globally to $x$ it also converges discretely. Now

$$r_n(\lambda I + L)x = (\lambda I_n + L_n)x_n + r_n(\lambda I + L)x - (\lambda I_n + L_n)x_n$$
$$= (\lambda I_n + L_n)x_n + r_n(\lambda I + L)x - (\lambda I_n + L_n)r_n x$$
$$+ (\lambda I_n + L_n)(r_n x - x_n).$$

From the consistency and uniform boundedness of the $L_n$ and the fact that $\|r_n x - x_n\| \to 0$ it follows that

$$\lim_{n \to \infty} \|r_n(\lambda I + L)x - (\lambda I_n + L_n)x_n\| = 0.$$

Applying now (4.40) completes the proof. ∎

**THEOREM 6.22.** Let $\{k_n\}$ be a collectively compact sequence of operators $X_n \to X$, and let $\{l_n\}$ be a sequence of linear operators $X \to X$, uniformly bounded and such that, for every $z \in X, l_n z$ converges to an element in $X$. Then the sequence $\{l_n k_n\}$ is collectively compact.

PROOF. Since $\{k_n\}$ is collectively compact every bounded sequence $\{x_n\}$ contains a subsequence $\{x_{n'}\}$ such that $\{k_{n'} x_{n'}\}$ converges to some $z \in X$. Now if $x_{n'}$ and $x_{n''}$ denote elements of this subsequence, then

$$\|l_{n'} k_{n'} x_{n'} - l_{n''} k_{n'} x_{n''}\| \le \|l_{n'}\| \|k_{n'} x_{n'} - z\| + \|l_{n''}\| \|k_{n''} x_{n''} - z\|$$
$$+ \|(l_{n'} - l_{n''})z\|.$$

For sufficiently large $n'$ and $n''$ the right-hand side can be made as small as desired. Therefore $\{l_{n'} k_{n'} x_{n'}\}$ is a Cauchy sequence which, because of the completeness of $X$, converges to an element in $X$. ∎

We are now ready to prove the main stability theorem.

**THEOREM 6.23.** If

(a) $\{A_n, r_n, p_n\}$ is a stable and consistent approximation of $A$;
(b) $\{K_n, r_n, p_n\}$ is a consistent approximation of $K$;
(c) $\{K_n\}$ is uniformly bounded;
(d) $\{p_n K_n\}$ is collectively compact;
then $\{\lambda A_n + K_n, r_n, p_n\}$ is a stable and consistent approximation of $(\lambda A + K)$.

PROOF.   Consistency follows easily from the assumptions. Assume now that $\{\lambda A_n + K_n\}$ is not stable. This implies that there exists a sequence $\{x_n\}$ with $\|x_n\| = 1$ such that

$$(\lambda A_n + K_n)x_n = \epsilon_n, \tag{6.36}$$

with

$$\lim_{n \to \infty} \|\epsilon_n\| = 0. \tag{6.37}$$

Since $\{A_n\}$ is stable (6.36) is equivalent to

$$\left(\lambda I_n + A_n^{-1}K_n\right)x_n = A_n^{-1}\epsilon_n. \tag{6.38}$$

For any $x \in X$ we have

$$A^{-1}x - p_n A_n^{-1} r_n x = p_n\left[r_n A^{-1}x - A_n^{-1}r_n x\right] - p_n r_n A^{-1}x + A^{-1}x,$$

so that $p_n A_n^{-1} r_n x \to A^{-1}x$. Applying Theorem 6.22 with $l_n = p_n A_n^{-1} r_n$ and $k_n = p_n K_n$ we see that $\{p_n A_n^{-1} r_n p_n K_n\}$ is collectively compact. But

$$p_n A_n^{-1} r_n p_n K_n = p_n A_n^{-1} K_n,$$

so that there exists a subsequence $\{x_{n'}\}$ such that

$$p_{n'} A_{n'}^{-1} K_{n'} x_{n'} \to -\lambda \bar{x} \in X.$$

From (6.38)

$$\lambda p_{n'} x_{n'} = p_{n'} A_{n'}^{-1} \epsilon_{n'} - p_{n'} A_{n'}^{-1} K_{n'} x_{n'},$$

so $p_{n'} x_{n'} \to \bar{x}$. Applying now Theorem 6.21 we get

$$\lim_{n' \to \infty} \left\|\left(\lambda I_{n'} + A_{n'}^{-1} K_{n'}\right)x_{n'}\right\| = \left\|(\lambda I + A^{-1}K)\bar{x}\right\|. \tag{6.39}$$

By (6.38) the left-hand side of this expression is zero and (6.39) implies that there exists an $\bar{x}$, with $\|\bar{x}\| \neq 0$, such that

$$(\lambda I + A^{-1}K)\bar{x} = 0.$$

This contradicts the assumption that our equation is uniquely solvable and the theorem is proved.  ■

This theorem represents a generalization of the stability results derived in Sections 6.2–6.4, since for integral equations $A_n = I_n$, making condition (a) of Theorem 6.23 trivial. It should also be noted that condition (d) of the theorem is closely related, but somewhat more general than (6.30). It is easily shown that if (6.30) is satisfied, then $\{p_n K_n\}$ is necessarily collectively compact (Exercise 3).

To prove stability in more general cases we need to verify conditions (a)–(d) of Theorem 6.23. Conditions (b) and (c) are simple and are satisfied for most reasonable methods. Condition (a) requires separate consideration, but we assumed that this could be done. We are then left with verifying (d); as already remarked before we only need to find some appropriate $\{p_n\}$, and there are usually many choices. A theorem giving sufficient conditions on $p_n$ can be found in (Linz 1977), but we do not go into detail here. The next example should give the reader a sufficiently good idea on how to proceed in specific cases.

**Example 6.4.**   Consider the integro-differential equation

$$u''(s) - u(s) + \int_0^1 k(s,t)u(t)\,dt = g(s) \tag{6.40}$$

$$u(0) = u(1) = 0.$$

To discretize we proceed in the usual fashion and introduce the uniform partition $0 \le t_{n1} < t_{n2} < \cdots < t_{nn} = 1, t_{n,i+1} - t_{ni} = h$. The second derivative is replaced by a centered difference and the integral by the trapezoidal rule. The approximate solution $\{U_i\}$ is then determined by the system

$$\frac{U_{i-1} - (2+h^2)U_i + U_{i+1}}{h^2} + h\sum_{j=1}^{n}{}' k(t_{ni},t_{nj})U_j = g(t_{ni}), \tag{6.41}$$

where the prime in the summation is used to indicate that the first and last terms are to be halved. The consistency and stability of the differential part of the operator has already been established in Example 4.6. The consistency and uniform boundedness of the integral part are obvious and we need only verify condition (d). Since the underlying space $X$ is $C[0,1]$ and $X_n$ is $R^n$ the piecewise linear interpolation described in Example 6.3 can be used for $p_n$. If $\{\mathbf{x}_n\}$ is any bounded sequence, then

$$(K_n\mathbf{x}_n)_i = h\sum_{j=0}^{n}{}' k(t_{ni},t_{nj})x_{nj},$$

$$(p_n K_n\mathbf{x}_n)(s) = h\sum_{i=1}^{n} \varphi_{ni}(s)\sum_{j=1}^{n}{}' k(t_{ni},t_{nj})x_{nj},$$

and we immediately note that the functions $p_n K_n x_n$ are uniformly bounded. Also

$$(p_n K_n x_n)(s_1) - (p_n K_n x_n)(s_2) = h \sum_{i=1}^{n} \left[ \varphi_{ni}(s_1) - \varphi_{ni}(s_2) \right] \sum_{j=1}^{n} {}' k(t_{ni}, t_{nj}) x_{nj}$$

$$= h \sum_{j=1}^{n} {}' x_{nj} \sum_{i=1}^{n} \left[ \varphi_{ni}(s_1) - \varphi_{ni}(s_2) \right] k(t_{ni}, t_{nj}).$$

Since

$$\sum_{i=1}^{n} \varphi_{ni}(s) k(t_{ni}, t_{nj})$$

is the piecewise linear interpolant for $k(s, t_{nj})$ it follows by a simple argument that the set of functions $p_n K_n x_n$ is also equicontinuous and hence by the Arzela–Ascoli theorem it is compact. Therefore $\{p_n K_n\}$ is collectively compact and (6.41) is a stable, consistent and hence convergent approximation method. Under suitable smoothness assumptions on $u(s)$ and $k(s, t)$ the order of convergence is two (Exercise 5).

### Exercises 6.5

1. Prove Theorem 6.20.
2. In Theorem 6.23 show that the consistency of $\{\lambda A_n + K_n\}$ follows from the assumptions.
3. Show that if (6.30) is satisfied, then $\{p_n K_n\}$ is collectively compact.
4. Complete the argument in Example 6.4 claiming that the set $\{p_n K_n x_n\}$ is equicontinuous.
5. Investigate the order of convergence of the method in Example 6.4.

# 7

# THE FINITE ELEMENT METHOD

In recent years a class of methods, termed *finite element methods*, has become very prominent in the literature of numerical analysis. The approach was first introduced by engineers whose motivation was primarily an intuitive one and for some time the rigorous analysis lagged behind the practical applications. Currently, however, a considerable effort is being devoted to closing this gap, although the development is by no means complete. In this chapter we try to single out some of the general ideas characterizing finite element methods, omitting as far as possible a multitude of rather difficult technical details. For more information the reader may consult (Strang and Fix, 1973), (Schultz, 1972), and (Prenter, 1975).

## 7.1 VARIATIONAL FORMULATION OF OPERATOR EQUATIONS AND GALERKIN'S METHOD

We have paid considerable attention to the linear operator equation

$$Lx = y. \tag{7.1}$$

This general form is basic to numerical analysis since the representation of physical systems as mathematical equations, such as differential and integral equations, is a common and familiar process. It is not the only way though, and often it is possible (and perhaps more intuitive) to seek the solution as that element of a linear space which minimizes a certain functional (e.g., the total energy of a system). Variational calculus deals with analytical methods for solving problems formulated this way; here we are concerned only with the use of the variational approach for numerical purposes. We begin by noting that under certain circumstances the operator equation (7.1) has a simple variational equivalent.

**THEOREM 7.1.** Let $X$ and $Y$ be subspaces of an inner product space $Z$, and let $L:X \to Y$ be a linear, positive-definite, symmetric operator. If $x$ satisfies (7.1), then it minimizes the functional

$$J(x) = (Lx,x) - 2(y,x) \qquad (7.2)$$

that is,

$$J(x) \leq J(\bar{x}), \qquad \text{for all } \bar{x} \in X.$$

PROOF. Let $\bar{x} = x + \Delta x$. Then

$$\begin{aligned}
J(\bar{x}) &= J(x + \Delta x) \\
&= (L(x + \Delta x), x + \Delta x) - 2(y, x + \Delta x) \\
&= (Lx,x) + 2(Lx,\Delta x) + (L\Delta x,\Delta x) - 2(y,x) - 2(y,\Delta x) \\
&= J(x) - 2(Lx - y, \Delta x) + (L\Delta x,\Delta x) \\
&= J(x) + (L\Delta x,\Delta x).
\end{aligned}$$

But since $L$ is positive-definite we have $(L\Delta x,\Delta x) \geq 0$, completing the proof. ∎

**THEOREM 7.2.** If the assumptions of the previous theorem are satisfied and if $x$ minimizes $J$ over $X$, then $x$ satisfies

$$(Lx - y, z) = 0, \qquad \text{for all } z \in X. \qquad (7.3)$$

PROOF.

$$J(x + \Delta x) = J(x) + 2(Lx,\Delta x) - 2(y,\Delta x) + (L\Delta x,\Delta x).$$

Since $x$ minimizes $J$ we must have for all $\Delta x \in X$

$$2(Lx,\Delta x) - 2(y,\Delta x) + (L\Delta x,\Delta x) \geq 0.$$

If there were a $\Delta x$ such that $(Lx - y, \Delta x) \neq 0$, then the above expression could be made negative (why?). Therefore (7.3) must be satisfied. ∎

Equation (7.3) does not always imply that (7.1) is satisfied; all it says is that $Lx - y$ is orthogonal to $X$. Whether this implies $Lx = y$ depends on the nature of $X$ and $Y$. For the cases of practical interest this is so, and we will assume that if (7.3) is satisfied then (7.1) holds.

Thus minimizing the functional $J(x)$ provides an alternate, but essentially equivalent way of solving certain operator equations. Now in practice we are usually no more successful in minimizing $J(x)$ than we are in obtaining a closed form solution of (7.1). We must therefore rely on approximate methods which we construct by minimizing $J(x)$ not over all of $X$, but only over a finite-dimensional subspace thereof.

Let us choose a set of linearly independent elements $\varphi_{ni}, i = 1, 2, \ldots, n$, $n = 1, 2, \ldots$ in $X$ and minimize $J(x)$ over $X_n = \text{span} \{\varphi_{ni}\}$. Then we can write

$$x_n = \sum_{i=1}^{n} \alpha_i \varphi_{ni}$$

and

$$J(x_n) = \sum_{i=1}^{n} \sum_{j=1}^{n} \alpha_i \alpha_j \left(L\varphi_{ni}, \varphi_{nj}\right) - 2 \sum_{i=1}^{n} \alpha_i (\varphi_{ni}, y).$$

Since $J(x_n)$ is a quadratic function of the $\alpha_i$ we can find the minimum by differentiating with respect to $\alpha_i$ and setting the result to zero. This yields the linear system

$$\sum_{j=1}^{n} \alpha_j \left(L\varphi_{ni}, \varphi_{nj}\right) = (\varphi_{ni}, y), \qquad i = 1, 2, \ldots, n. \tag{7.4}$$

We recognize this as Galerkin's method which we have already encountered in Chapter 4. (For positive-definite $L$ the method is often called the Rayleigh–Ritz method.)

It would seem then that we have accomplished nothing by introducing the variational form, since our result was just as easily arrived at starting from the operator form of the equation. But there is a significant difference: the functional $J(x)$ may be defined (by an appropriate extension) not only for $x \in X$, but also for elements in a larger space and hence we can look for solutions and trial functions in this larger space. Any $x$ for which $J(x)$ (in its extended form) is minimized is called a *generalized solution*. Since we shall assume that there exists a unique solution in $X$ we shall not be concerned with the computation of a generalized solution, but we shall make use of the fact that the expansion functions are not necessarily in $X$.

To make this idea a bit more precise let us consider the problem from a somewhat different point of view. For $L$ symmetric and positive-definite the expression

$$(x_1, x_2)_L = (Lx_1, x_2) \tag{7.5}$$

defines an inner product on $X$, as is easily verified. This is the so-called *energy inner product*; the corresponding norm is the *energy norm*. With this inner product the space $X$ is then another inner product space, but it is not necessarily complete. By completing $X$ with respect to the energy norm we obtain a Hilbert space $H_L$, which we call the *energy space*. For elements in $H_L$ but not in $X$, the energy inner product may not be given by (7.5) since $L$ may not be defined for all of $H_L$. What we can try to do is to redefine $(,)_L$ in such a way that it is defined for all of $H_L$ and so that it reduces to $(Lx_1, x_2)$ when $x_1$ and $x_2$ are in $X$. This is what is meant by "extending" the inner product (or, more generally, any operator).

The norms in $X$ and $H_L$ are generally not equivalent, but we do have Theorem 7.3.

**THEOREM 7.3.** If $\|x_n - x\|_L \to 0$, then $\|x_n - x\| \to 0$.

**PROOF.**

$$\|x_n - x\|_L^2 = (x_n - x, x_n - x)_L$$
$$= (L(x_n - x), x_n - x).$$

Since $L$ is positive-definite we have by Definition 4.2

$$\|x_n - x\|_L^2 \geq \gamma \|x_n - x\|^2, \qquad \gamma > 0,$$

so that

$$\|x_n - x\| \leq \frac{1}{\sqrt{\gamma}} \|x_n - x\|_L.$$

On the other hand, $\|x_n - x\| \to 0$ does not generally imply convergence in the energy norm unless $L$ is bounded. ∎

If we let $(,)_L$ denote the extension of the energy inner product to $H_L$, then

$$J(x) = (x, x)_L - 2(x, y)$$

defines the extended functional. We now propose to solve our problem by minimizing $J(x)$ over $H_L$. This immediately raises the question whether the answer we so obtain is still the solution of (7.1) or whether enlarging the space has introduced a spurious minimum.

**THEOREM 7.4.** Let $x$ be the solution of (7.1). Then

$$J(x) \leq J(\bar{x}), \qquad \text{for all } \bar{x} \in H_L.$$

PROOF. We have already shown that $x$ minimizes $J$ over $X$. Assume now that there exists an $\bar{x} \in H_L$ but not in $X$ such that

$$J(\bar{x}) < J(x). \tag{7.6}$$

Since $H_L$ is the completion of $X$ in the energy norm there exists an $x^* \in X$ arbitrarily close to $\bar{x}$, that is, for any $\epsilon > 0$ there is an $x^*$ such that

$$\|\bar{x} - x^*\|_L \le \epsilon.$$

But

$$
\begin{aligned}
J(\bar{x}) &= J(x^* + (\bar{x} - x^*)) \\
&= (x^* + (\bar{x} - x^*), x^* + (\bar{x} - x^*))_L - 2(y, x^* + (\bar{x} - x^*)) \\
&= J(x^*) + 2(\bar{x} - x^*, x^*)_L + (\bar{x} - x^*, \bar{x} - x^*)_L - 2(y, \bar{x} - x^*).
\end{aligned}
$$

Then

$$|J(\bar{x}) - J(x^*)| \le 2\|\bar{x} - x^*\|_L \|x^*\|_L + \|\bar{x} - x^*\|_L^2 + 2\|y\| \|\bar{x} - x^*\|$$

and from Theorem 7.3

$$|J(\bar{x}) - J(x^*)| \le 2\|\bar{x} - x^*\|_L \|x^*\|_L + \|\bar{x} - x^*\|_L^2 + \frac{2\|y\|}{\sqrt{\gamma}} \|\bar{x} - x^*\|_L.$$

This implies that there exists an $x^* \in X$ such that

$$|J(x^*) - J(\bar{x})| \le c\epsilon$$

for arbitrarily small $\epsilon$, contradicting the assumption that $x$ minimizes $J$ over $X$. ∎

This theorem tells us that in minimizing $J(x)$ we can use expansion functions from $H_L$ without changing the solution. This is a very useful observation since it allows the use of simpler $\varphi_{ni}$ and therefore facilitates the actual computations.

In the energy space the problem of minimizing $J(x)$ is just an approximation problem. With the energy inner product we can rewrite (7.4) as

$$\sum_{j=1}^{n} \alpha_j (\varphi_{ni}, \varphi_{nj})_L = (\varphi_{ni}, y)$$

$$= (\varphi_{ni}, x)_L. \tag{7.7}$$

But these are just the normal equations for the approximation of an element in an inner product space so that all the conclusions reached in Section 2.4 hold. In particular we have Theorem 7.5.

## THEOREM 7.5.

    (a)   There exists a unique $x_n \in X_n = \mathrm{span}\; \{\varphi_{n1}, \ldots, \varphi_{nn}\}$ which minimizes $J(x)$ over $X_n$.

    (b)   If the set $\{\varphi_{ni}\}$ is closed in $H_L$, then the sequence $\{x_n\}$ converges to $x$ in the energy norm.

    (c)   $(x - x_n, z)_L = 0$ for all $z \in X_n$.

PROOF.  Part (a) is essentially Theorem 2.11. Parts (b) and (c) follow from Theorem 2.17.  ∎

We have already encountered one theorem (Theorem 4.10) dealing with the convergence of Galerkin's method, but it is worth noting the difference between that theorem and the above.

In Theorem 4.10 we assumed that the set $\{\varphi_{ni}\}$ was closed in $X$ and that $L$ was bounded; in Theorem 7.5 we require that $\{\varphi_{ni}\}$ be closed in $H_L$, but make no assumption of boundedness on $L$. Another difference is that in Theorem 4.10 we proved convergence in the $X$-norm, while here we obtained convergence in the energy norm (which is a stronger result, according to Theorem 7.3).

While the method we obtain by starting from the variational formulation is formally Galerkin's method we have gained something because we can now take the expansion functions in the larger space $H_L$. Our analysis shows that this procedure produces approximations that converge to the true solution in the energy norm as well as the $X$-norm. Still, there remain a number of important questions which must be answered before we can claim any practical importance for this class of methods:

    1.   We have already remarked that a convergence proof without a guarantee of stability is of limited usefulness. Galerkin's method is not unconditionally stable, the stability depends on the choice of the expansion functions.

    2.   We have proved convergence in the energy norm, but in practice it is often desirable to be assured of convergence in a stronger norm, for example, the maximum norm. Rates of convergence are also needed.

    3.   The expansion functions should be chosen such that the method is stable, but this is not the only consideration. The $\varphi_{ni}$ should be

relatively simple so that $(L\varphi_{ni},\varphi_{nj})$ and $(\varphi_{ni},y)$ are easily computed. If possible, the resulting matrix should be sparse to facilitate the numerical computations.

To produce a general theory to deal with these problems is well beyond the scope of this simple discussion. More details can be found in the work already cited, as well as in (Mikhlin, 1971) and (Aubin, 1972) who give a thorough treatment of the theoretical aspects. Here we will be satisfied with elucidating the main points by presenting a specific example. We have chosen what is perhaps the simplest nontrivial example, the second-order linear differential equation

$$-\frac{d^2x(t)}{dt^2} + q(t)x(t) = y(t), \qquad 0 \le t \le 1 \qquad (7.8)$$

$$x(0) = x(1) = 0,$$

with $0<q(t)<Q$. In spite of the elementary nature of this problem the analysis shows fairly clearly the techniques and problems common to the more difficult cases.

## Exercises 7.1

1. Show that if there exists a $\Delta x$ such that $(Lx-y,\Delta x)\neq0$, then $2(Lx,\Delta x) -2(y,\Delta x)+(L\Delta x,\Delta x)$ can be made negative.
2. Show that if $L$ is bounded, then the norms $\| \|$ and $\| \|_L$ are equivalent.
3. Show that if $L:X\rightarrow X$, then (7.3) implies (7.1).
4. Show that the operator in (7.8) is symmetric and positive-definite.

## 7.2   THE CONSTRUCTION OF THE FINITE ELEMENTS

If we use the inner product

$$(x,y) = \int_0^1 x(t)y(t)dt, \qquad (7.9)$$

then the energy inner product corresponding to (7.8) is

$$(x,y)_L = -\int_0^1 \frac{d^2x(t)}{dt^2}y(t)dt + \int_0^1 q(t)x(t)y(t)dt. \qquad (7.10)$$

Integrating by parts and using the homogeneous boundary conditions we have

$$(x,y)_L = \int_0^1 x'(t)y'(t)\,dt + \int_0^1 q(t)x(t)y(t)\,dt. \qquad (7.11)$$

For $x,y \in C^{(2)}[0,1]$, (7.10) and (7.11) are equivalent, but (7.11) represents an extension of (7.10) since it is defined for all functions for which the integrals in (7.11) exist as Lebesgue integrals.

In choosing expansion functions for Galerkin's method we are then limited to functions for which the right-hand side of (7.11) exists in the Lebesgue sense, but it must be remembered that not all such functions are necessarily permitted. We must also be sure that the $\varphi_{ni}$ are elements of $H_L$; we are not particularly concerned here with the actual structure of $H_L$, but we must be able to show that the expansion functions are limits of certain Cauchy sequences in $C^{(2)}[0,1]$. Also, for practical reasons, some consideration should be given to the simplicity of the final numerical scheme.

As a start, let us construct a rather simple method by requiring that the expansion functions be piecewise linear. More specifically, let $\pi_n$ be the partition of $[0,1]$ generated by the distinct points $t_{n1}, t_{n2}, \ldots, t_{nn}$, $t_{ni} = (i-1)h, h = 1/(n-1)$. Then every piecewise linear function can be represented as a linear combination of the hat functions defined in Example 6.3,

$$\begin{aligned}
\varphi_{ni}(t) &= (t - t_{n,i-1})/h, & t_{n,i-1} \le t \le t_{ni}, \\
&= (t_{n,i+1} - t)/h, & t_{ni} \le t \le t_{n,i+1}, \\
&= 0, & \text{otherwise.}
\end{aligned}$$

A simple calculation yields the following matrix elements for Galerkin's method:

$$\begin{aligned}
(\varphi_{ni}, \varphi_{ni})_L &= \frac{1}{h^2} \int_{t_{n,i-1}}^{t_{ni}} \left[ 1 + q(t)(t - t_{n,i-1})^2 \right] dt \\
&\quad + \frac{1}{h^2} \int_{t_{ni}}^{t_{n,i+1}} \left[ 1 + q(t)(t_{n,i+1} - t)^2 \right] dt \\
(\varphi_{ni}, \varphi_{n,i-1})_L &= (\varphi_{n,i-1}, \varphi_{ni})_L \\
&= \frac{1}{h^2} \int_{t_{n,i-1}}^{t_{ni}} \left[ q(t)(t - t_{n,i-1})(t_{ni} - t) - 1 \right] dt \\
(\varphi_{ni}, \varphi_{nj})_L &= 0, \qquad \text{otherwise.} \qquad (7.12)
\end{aligned}$$

The matrix is tridiagonal which simplifies the numerical computations considerably.

We must now ask whether we are justified in using such expansion functions. Since the derivative of $\varphi_{ni}$ exists and is bounded almost everywhere the integrals in (7.11) exist. To show that $\varphi_{ni} \in H_L$ we can construct a Cauchy sequence in $C^{(2)}$ converging to $\varphi_{ni}$; the details of one particular construction are outlined in Exercise 1.

Next we might ask why one does not just simply use polynomial approximations. One reason lies in the fact that polynomials have undesirable properties as approximants in that they tend to produce unstable algorithms. Again, the simplest case is $L = I$, $\varphi_{ni}(t) = t^{i-1}$, leading to the Hilbert matrix. The use of piecewise polynomials avoids this stability problem as we will see in Section 7.4. Furthermore, the use of polynomials leads to a full (rather than a tridiagonal) matrix which decreases the efficiency of the computations. Therefore, piecewise polynomials are much more appropriate and it is this combination, Galerkin's method with piecewise polynomial approximants which is known as the finite element method.

We note that the basis $\varphi_{ni}$ for the expansion has the property that $\varphi_{ni}$ is nonzero only over an interval of width $O(h)$. This is the reason why the resulting matrix is sparse. Any basis whose elements vanish outside some interval of width $O(h)$ is called a *local* or *patch basis*.

To construct a local basis for piecewise quadratics is a somewhat more difficult but still manageable task. Assume that the interval $[0,1]$ is divided into $n$ equal subintervals and that we require that the approximating functions are quadratics in each subinterval. Every such function can be expressed as a linear combination of the three piecewise quadratics $\varphi_{ni}^{(M)}, \varphi_{ni}^{(L)}, \varphi_{ni}^{(R)}$, defined by (writing $t_i$ for $t_{ni}$)

$$\varphi_{ni}^{(M)}(t_i + h/2) = 1, \qquad \varphi_{ni}^{(M)}(t_i) = \varphi_{ni}^{(M)}(t_i + h) = 0$$

$$\varphi_{ni}^{(M)}(t) = 0, \qquad t \notin [t_i, t_i + h]$$

$$\varphi_{ni}^{(R)}(t_i) = 1, \qquad \varphi_{ni}^{(R)}(t_i + h/2) = \varphi_{ni}^{(R)}(t_i + h) = 0,$$

$$\varphi_{ni}^{(R)}(t) = 0, \qquad t \notin [t_i, t_i + h],$$

$$\varphi_{ni}^{(L)}(t_i + h) = 1, \qquad \varphi_{ni}^{(L)}(t_i) = \varphi_{ni}^{(L)}(t_i + h/2) = 0$$

$$\varphi_{ni}^{(L)}(t) = 0, \qquad t \notin [t_i, t_i + h].$$

The functions $\{\varphi_{ni}^{(M)}, \varphi_{ni}^{(R)}, \varphi_{ni}^{(L)}\}$ are then a basis for the space of all piecewise quadratics, but this space is actually too large since it contains functions which are not continuous. To remedy this situation we define a basis $\{\varphi_{ni}, \psi_{ni}\}$ by

$$\varphi_{ni} = \varphi_{ni}^{(M)}, \qquad i = 1, 2, \ldots, n,$$

$$\psi_{ni} = \varphi_{ni}^{(R)} + \varphi_{n,i-1}^{(L)}, \qquad i = 1, 2, \ldots, n+1, \qquad (7.13)$$

which span the space of all piecewise continuous quadratics. Expressions for the matrix elements for Galerkin's method require some simple manipulations which are left as an exercise.

As a final example let us consider approximations with cubic splines. The cardinal splines described in Section 2.3 form a possible basis, but it is not local and therefore undesirable for computations. A local basis is given by the so-called *B-splines* defined as follows: Assuming that $[0, 1]$ is divided into $n$ intervals of width $h$, we take

$$B_i(t) = \begin{cases} (t - t_{i-2})^3 / h^3, & t_{i-2} \le t \le t_{i-1} \\ 1 + 3(t - t_{i-1})/h + 3(t - t_{i-1})^2/h^2 - 3(t - t_{i-1})^3/h^3, & t_{i-1} \le t \le t_i \\ 1 + 3(t_{i+1} - t)/h + 3(t_{i+1} - t)^2/h^2 - 3(t_{i+1} - t)^3/h^3, & t_i \le t \le t_{i+1} \\ (t_{i+2} - t)^3/h^3, & t_{i+1} \le t \le t_{i+2}, \\ 0, & \text{otherwise.} \end{cases} \qquad (7.14)$$

This definition holds for $i = 0, 1, \ldots, n+1$ with $t_i = (i-1)h$. It is not difficult to show that the $B_i$ are cubic splines and that they form a basis for the space of cubic splines. The matrix for Galerkin's method is still sparse, although more complicated than in the preceding cases. There is of course a tradeoff here: if we choose more complicated elements we can achieve faster convergence, but this improvement comes at the expense of practical complication and a loss of computational efficiency.

### Exercises 7.2

1. Let $f(t)$ be the function

$$f(t) = |t|, \qquad -1 \le t \le 1.$$

Consider now the sequence of functions $f_n(t)$ defined by

$$f_n(t) = f(t), \qquad t \notin [-1/n, 1/n],$$

with $f_n(t)$ in $[-1/n, 1/n]$ being a fourth degree polynomial chosen so that $f_n \in C^{(2)}[-1, 1]$.

(a)  Show that $\sup_n \|f_n'\|_\infty \leq M < \infty$.
(b)  Use this to show that $\{f_n\}$ is a Cauchy sequence converging to $f$ in the energy norm defined by (7.11).
(c)  From this show that the piecewise linear functions are in $H_L$.

2.  Show that for piecewise constant functions the integrals in (7.11) exist, but that such elements are inappropriate as expansion functions.

3.  Prove that the $\varphi_{ni}$ in (7.13) are a basis for the space of piecewise quadratics.

4.  (a)  Show that the $B_i$ defined by (7.14) are cubic splines.
    (b)  Sketch a $B$-spline.
    (c)  Show that $B_0, \ldots, B_{n+1}$ span the space of cubic splines.

## 7.3  CONVERGENCE RATES FOR THE FINITE ELEMENT METHOD

In Theorem 7.5 we have established that the Galerkin solution $x_n$ converges to the true solution if the set $\{\varphi_{ni}\}$ is closed in $H_L$. We now elaborate on this by establishing convergence rates in various norms. From Theorem 7.5 we have that

$$\|x - x_n\|_L \leq \|x - z_n\|_L, \qquad \text{for all } z_n \in X_n. \tag{7.15}$$

Therefore, to bound the error we only need to find a bound for some $z_n$; the error in the Galerkin solution cannot be larger.

Returning now to the specific problem (7.8) and taking the hat functions as basis we can use approximation theory to bound $\|x - z_n\|_L$. Take as $z_n$ the linear interpolating function to $x$, with $z_n(0) = z_n(1) = 0$; then $z_n \in X_n$. If we assume that $x \in C^{(2)}[0, 1]$ with $\|x''\|_\infty \leq M_2$, then

$$\|x - z_n\| \leq \frac{h^2}{8} M_2.$$

To compute a bound for $\|x - z_n\|_L$ we also need a bound on $|x' - z_n'|$. From Exercise 8, Section 2.2 we see that

$$\sup_{t_{i+1} < t < t_i} |x'(t) - z_n'(t)| \leq hM_2,$$

so that

$$\|x - z_n\|_L^2 = \int_0^1 (x'(t) - z_n'(t))^2 \, dt + \int_0^1 q(t)(x(t) - z_n(t))^2 \, dt,$$

$$\|x - z_n\|_L^2 \leq \left(h^2 + \frac{1}{64} Qh^4\right) M_2^2,$$

$$\|x - z_n\|_L \leq ChM_2. \tag{7.16}$$

This result must be considered somewhat disappointing since it only establishes first-order convergence while a piecewise linear approximation is capable of giving second-order accuracy. We must therefore ask whether the convergence is in fact second order and if so how we can modify our analysis to prove it. The trick here lies in noticing that the derivatives of the solution can be bounded in terms of the right-hand side of the equation.

**THEOREM 7.6.** Let $f \in C[0, 1]$, and let $z$ be the solution of (7.8) with $f$ as the right-hand side. Then there exists a $\gamma < \infty$, independent of $f$, such that

$$\|z''\|_\infty \leq \gamma \|f\|_\infty. \tag{7.17}$$

PROOF. The operator $-d^2/dt^2[\quad] + q(t)[\quad]$ with homogeneous boundary conditions has a bounded inverse when considered as a mapping $C^{(2)}[0, 1] \to C[0, 1]$ with maximum norm. Hence, for some $\gamma_1$,

$$\|z\|_\infty \leq \gamma_1 \|f\|_\infty.$$

But

$$z'' = qz - f,$$

so that

$$\|z''\|_\infty \leq Q\gamma_1 \|f\|_\infty + \|f\|_\infty \leq \gamma \|f\|_\infty \tag{7.18}$$

∎

In this way the second derivative of the solution can be bounded in terms of the right-hand side of the equation. In general, a problem in which the derivatives can be so bounded is said to be *strongly coercive*, a concept which plays an important role in the error analysis of finite element methods.

Returning now to the original equation we see from (7.16) and (7.18) that

$$\|x - x_n\|_L \leq C_1 h \|y\|_\infty. \tag{7.19}$$

We can also use the coerciveness to improve the error bound.

**THEOREM 7.7.** If $x$ is the solution of (7.8) and $x_n$ is its Galerkin approximation using piecewise linear approximants, then

$$\|x - x_n\| = O(h^2). \tag{7.20}$$

PROOF. In (7.8) take $e_n = x - x_n$ as the right-hand side and call the resulting solution $z$. Then

$$(z,v)_L = (e_n,v), \qquad \text{for all } v \in X.$$

Putting $v = e_n$ we have

$$(z,e_n)_L = (e_n,e_n) = \|e_n\|^2.$$

From Theorem 7.5, part (c)

$$(e_n,v_n)_L = 0, \qquad \text{for all } v_n \in X_n,$$

so that

$$(z - v_n, e_n)_L = \|e_n\|^2$$

and, by the Schwarz inequality

$$|(z - v_n, e_n)_L| \leq \|z - v_n\|_L \|e_n\|_L.$$

If we now choose the Galerkin approximation to $z$ as $v_n$, then from (7.19) there exists a $C_2$ such that

$$\|z - v_n\|_L \leq C_2 h \|e_n\|_\infty$$
$$\leq C_2 h \|e_n\|.$$

Putting these bounds together yields

$$\|e_n\|^2 \leq C_2 h \|e_n\| \|e_n\|_L$$
$$\leq C_1 C_2 h^2 \|e_n\| \|y\|_\infty,$$

establishing (7.20). ∎

Finally, let us consider the question of convergence in the maximum norm. Convergence in the inner product norm proved above is of course not sufficient to guarantee convergence in the maximum norm. To prove the stronger result we must again rely on suitable manipulation of the problem.

If $z(t)$ is a continuous and piecewise differentiable function with $z(0) = 0$, then

$$z(t) = \int_0^t z'(s)\,ds.$$

Using Schwarz's inequality we see that

$$|z(t)| \leq \left( \int_0^t (z'(t))^2 \, dt \right)^{1/2},$$

or

$$\|z\|_\infty \leq \|z'\|.$$

Setting $z = x - x_n$ we get

$$\|x - x_n\|_\infty \leq \|x' - x_n'\|. \tag{7.21}$$

Now

$$\|x - x_n\|_L^2 = \int_0^1 (x' - x_n')^2 \, dt + \int_0^1 q(x - x_n)^2 dt,$$

and since $q(t) > 0$,

$$\|x' - x_n'\|^2 \leq \|x - x_n\|_L^2.$$

Therefore, from (7.19) and (7.21)

$$\|x - x_n\|_\infty \leq C_1 h \|y\|_\infty,$$

showing that the approximate solution converges to the true solution in the maximum norm.

## 7.4   STABILITY OF THE FINITE ELEMENT METHOD

At this point one might raise some objections to the procedure outlined in the previous section. It is perhaps not clear why one needs to proceed as we have done; Galerkin's method is after all just a special case of the general theory developed in Chapter 4. It would therefore seem appropriate to simply prove consistency and stability for the method. One reason for treating the problem in the way we did is simply that this analysis seems to be preferred by workers in the field and it is necessary to know their arguments in order to appreciate their work. There is, however, a more persuasive reason. Finite element methods are generally not stable in the sense of Definition 4.6, but only weakly stable. The results obtainable from a consistency-stability analysis consequently tend to be less satisfactory than those resulting from variational arguments.

To explore the stability problem, let us further simplify problem (7.8) by taking $q(t)=0$. Using the hat functions as finite elements the matrix for Galerkin's method in an $(n-1)\times(n-1)$ matrix of the form

$$L_n = \frac{1}{h}\begin{bmatrix} 2 & -1 & & \\ -1 & 2 & -1 & \\ & -1 & \ddots & \\ & & \ddots & \ddots \\ & & & -1 & 2 \end{bmatrix}. \qquad (7.22)$$

This matrix is simple enough to have known eigenvalues which are

$$\lambda_i = \frac{2}{h}\left(1-\cos\frac{i\pi}{n}\right), \qquad i = 1,2,\ldots,n-1.$$

If $\lambda_{max}$ and $\lambda_{min}$ denote the largest and smallest eigenvalues, then

$$\lambda_{max} = O(n), \quad \lambda_{min} = O(1/n).$$

Since the matrix is symmetric we see from this that

$$\|L_n\|_2 = \lambda_{max} = O(n)$$
$$\|L_n^{-1}\|_2 = 1/\lambda_{min} = O(n).$$

The method is therefore weakly stable of degree one.

What does this imply about the numerical performance of the finite element method? We have already proved convergence, so the only possible source of difficulty can be the fact that (7.4) cannot be solved exactly. The two major contributing factors to this are (a) we often cannot evaluate the matrix elements $(\varphi_{ni},\varphi_{nj})_L$ and the right-hand side exactly, but must rely in numerical approximations, and (b) round-off error will affect the solution of the matrix equation. We can think of the contribution from both of these sources as a perturbation of size $\epsilon$ in the matrix elements. For the finite element method with a local basis the matrix is generally multiplied by $1/h$ as in (7.22); thus the effect on the matrix is of order $\epsilon/h$. This will be magnified, due to the weak stability, by another factor of $1/h$, so that the total effect is of order $\epsilon/h^2$. If $h$ is not too small, say larger than $10^{-2}$ and $\epsilon$ can be kept small, we can get respectable accuracy. If, however, high accuracy is needed we cannot achieve it by simply making $h$ small. A more acceptable way is to use more complicated finite elements, such as cubic splines, so that a small consistency error is obtainable with reasonably large $h$. It is known that the use of these more complicated elements will not make the stability problem significantly worse (Strang and Fix, 1973, Ch. 5). Thus, using higher-degree finite elements will decrease the discretization error quickly while keeping the effect of the computational errors within reasonable bounds.

Another way of describing the behavior of the finite element method is through the condition number of the associated matrix. Since round-off is essentially a relative perturbation it is proper to use the relative condition number

$$k(L_n) = \|L_n^{-1}\| \|L_n\|.$$

For symmetric matrices this is

$$k(L_n) = \lambda_{\max}/\lambda_{\min},$$

which is the traditional condition number used. For (7.22) this, of course, gives

$$k(L_n) = O(1/h^2),$$

leading to the same qualitative conclusions as in the preceding paragraph.

The brevity of treatment which we have accorded to the finite element method is not to be taken as a measure of its actual importance. The method is currently very popular and much effort is devoted to its study. The techniques used in this are similar to what we have outlined, but the practical problems encountered are considerably more difficult than the simple cases we looked at would indicate. The primary application of the finite element method is in the approximate solution of partial differential equations in two or more dimensions. There the questions of the proper choice of local bases, treatment of irregular shapes and boundary conditions, and solution of the resulting systems are quite complicated. The theoretical questions are equally difficult and while significant progress has been made, much remains to be done.

# 8

# THE SOLUTION OF NONLINEAR OPERATOR EQUATIONS BY DISCRETIZATION

The methods for solving the nonlinear inverse problem presented in Chapter 5 all involved replacing the original equation by a sequence of direct or linear problems. The discussion of convergence therefore centered around the question of the relation between the sequence of solutions so generated and the true solution. Implicit in that discussion was the assumption that each of the simpler problems can be solved and we ignored the fact that each step may involve further approximations. For example, if we wish to solve a nonlinear integral equation by Newton's method, then each step requires the solution of a linear integral equation. Since normally this cannot be done explicitly it becomes necessary to discretize the linear problems, for instance, by Nyström's method.

Perhaps a more direct approach is to discretize the nonlinear equation itself. This generates a system of nonlinear algebraic equations which must then be solved computationally. In such a procedure we must be concerned with the problem of convergence with respect to the discretization parameter, and the questions raised in Chapter 4 for the linear case must be reexamined. Some comments on this were already made in Section 5.3, here we will take a look at the notion of stability and the existence of asymptotic error expansions for nonlinear discretization techniques.

For the purpose of discussion we will consider the nonlinear equation in the form

$$Px = 0, \qquad (8.1)$$

with $P$ being an operator from a Banach space $X$ into a Banach space $Y$.

## 8.1 WELL-POSEDNESS AND CONDITION NUMBERS

In order to make any progress we must restrict the type of problem to be considered. In the linear case $Lx = y$ we assumed that $L$ had a bounded inverse which, in a Banach space setting, implies that the solution is unique and that bounded perturbations in the equation generate bounded perturbations in the solution. Thus we considered only well-posed problems. In the nonlinear case this characterization of well-posedness is too restrictive. In particular we need to give up the requirement of unique solvability since there are many nonlinear problems of practical interest having several solutions. (A trivial example is the solution of a quadratic equation with distinct roots.) Still, we want to retain the requirement of insensitivity of the solutions to small perturbations.

**DEFINITION 8.1.** Let $Z$ denote the solution set of (8.1), that is $Z = \{z \in X \,|\, Pz = 0\}$. Then (8.1) is said to be well-posed at $x \in Z$ if there exist constants $c > 0$, $\rho > 0$ such that

$$\|x_1 - x_2\| \le c\|P(x_1) - P(x_2)\| \tag{8.2}$$

for all $x_1, x_2 \in b(x, \rho)$.

If (8.2) is satisfied for all $x \in Z$, and if in addition there exist constants $k$ and $\epsilon > 0$ such that for every solution $x_\epsilon$ of $\|Px\| \le \epsilon$ there exists an $x \in Z$ such that

$$\|x - x_\epsilon\| \le k\epsilon \tag{8.3}$$

then we say that (8.1) is well-posed everywhere.

For simplicity it is convenient to replace (8.2) by an essentially equivalent condition on the Fréchet derivative of $P$.

**THEOREM 8.1.** Assume that $P$ is Fréchet differentiable in some neighborhood of $x \in Z$. Then (8.1) is well-posed at $x$ if and only if $P'(\bar{x})$ has a bounded inverse for all $\bar{x}$ near $x$.

**PROOF.** Assume that $P'(x_1)$ has a bounded inverse for $x_1$ near $x$. Then by the definition of the derivative

$$P'(x_1)(x_2 - x_1) = Px_2 - Px_1 + \eta(x_1, x_2) \tag{8.4}$$

where

$$\lim_{x_2 \to x_1} \frac{\|\eta(x_1, x_2)\|}{\|x_2 - x_1\|} = 0.$$

Therefore

$$\|x_2 - x_1\| \le \|P'(x_1)^{-1}\|[\|Px_2 - Px_1\| + \|\eta(x_1, x_2)\|],$$

or

$$\|x_2 - x_1\|\left[1 - \frac{\|P'(x_1)^{-1}\|\,\|\eta(x_1, x_2)\|}{\|x_2 - x_1\|}\right] \le \|P'(x_1)^{-1}\|\,\|Px_1 - Px_2\|.$$

The second term in the brackets on the left-hand side can be made as small as desired by taking $x_2$ sufficiently close to $x_1$. Thus

$$\|x_1 - x_2\| \le c\|Px_1 - Px_2\|.$$

Conversely, since we assumed that $P$ was differentiable near $x$, (8.4) will be satisfied near $x$. Therefore

$$\|P'(x_1)(x_1 - x_2)\| \ge \|Px_1 - Px_2\| - \|\eta(x_1, x_2)\|.$$

Applying (8.2) and making $\|\eta(x_1, x_2)\|$ small, we have

$$\|P'(x_1)(x_1 - x_2)\| \ge \left(\frac{1}{c} - \frac{\|\eta(x_1, x_2)\|}{\|x_1 - x_2\|}\right)\|x_1 - x_2\|.$$

By Theorem 4.7 and the fact that $P'(x_1)$ is linear it follows that $P'(x_1)$ has a bounded inverse. ∎

While well-posedness does not imply that (8.1) has a unique solution, it does guarantee local uniqueness, that is, that there is only one solution in some neighborhood of $x \in Z$ (Exercise 1). It also guarantees that the solutions are not overly sensitive to small perturbations.

**THEOREM 8.2.** Assume that $P$ is twice differentiable near $x \in Z$ and that (8.1) is well-posed at $x$. Let $Z_\epsilon$ be the set of solutions of

$$P\hat{x} = \epsilon. \tag{8.5}$$

Then, provided $\|\epsilon\|$ is sufficiently small, for every $x \in Z$ there exists an $\hat{x} \in Z_\epsilon$ such that

$$\|x - \hat{x}\| \le c\|P'(x)^{-1}\|\,\|\epsilon\|. \tag{8.6}$$

PROOF. First we must show that (8.5) has a solution near $x$. To do so consider the operator $G$ defined by

$$Gz = z - P'(x)^{-1}[Pz - \epsilon].$$

Then for $z_1$ and $z_2$ near $x$ we have

$$Gz_1 - Gz_2 = z_1 - z_2 - P'(x)^{-1}[Pz_1 - Pz_2]$$
$$= P'(x)^{-1}[P'(x) - P'(z_1)](z_1 - z_2) + P'(x)^{-1}\eta(z_1, z_2).$$

Since $P''(x)$ is assumed to exist, we can make $\|P'(x) - P'(z_1)\|$ as small as desired by taking $z_1$ close to $x$. Thus we get

$$\|Gz_1 - Gz_2\| \le \theta\|z_1 - z_2\|.$$

where $\theta$ is as small as we like. Since

$$Gx - x = P'(x)^{-1}\epsilon$$

we can apply the contraction mapping theorem in some neighborhood of $x$ showing that there exists a unique $\hat{z} \in b(x, r)$ such that

$$\hat{z} - P'(x)^{-1}[P\hat{z} - \epsilon] = \hat{z},$$

or

$$P\hat{z} = \epsilon.$$

The inequality (8.6) then follows by a simple expansion argument similar to that used in the proof of Theorem 8.1.     ■

Note that this theorem does not say that every solution of (8.5) is close to some solution of (8.1). To get such a result we need stronger conditions.

**THEOREM 8.3.** If $P$ is sufficiently differentiable and if (8.1) is well-posed everywhere, then, for sufficiently small $\epsilon$, for every $\hat{x} \in Z_\epsilon$ there exists an $x \in Z$ such that (8.6) is satisfied.

PROOF. That there exists such an $x \in Z$ is a direct consequence of the assumption that (8.1) is well-posed everywhere. The rest then follows from an expansion argument.     ■

In order to assure that small perturbations do not introduce any spurious solutions we have to assume that the problem is well-posed everywhere. The necessity of this assumption is demonstrated by the following example.

**Example 8.1.** The equation

$$e^t = 0$$

has no real solution. The problem is not well-posed everywhere since (8.3) is not satisfied. The perturbed problem

$$e^t = \epsilon$$

has a real solution for every $\epsilon > 0$.

The bound (8.6) indicates that $\|P'(x)^{-1}\|$ is a measure of the sensitivity of the solution to small perturbations. This quantity also appears when we compute the effect of a perturbation on $P$ (Exercise 2). Thus $\|P'(x)^{-1}\|$ is an appropriate choice for the condition number; since it depends on $x$ some of the solutions may be well-conditioned while others are not. This is of course analogous to the familiar example of the root-finding problem $f(x) = 0$, where $1/f'(x)$ is usually used as a condition number.

### Exercises 8.1

1. If (8.1) is well-posed at $x \in Z$, show that there exists some neighborhood $b(x,r)$ such that $x$ is the only solution in this neighborhood.
2. Let $Px = 0$ and assume that all conditions of Theorem 8.2 are satisfied. Let $\Delta P$ be a bounded perturbation of $P$.
   (a) Show that $(P + \Delta P)\hat{x} = 0$ has a solution for sufficiently small $\|\Delta P\|$.
   (b) Find a bound for $\|x - \hat{x}\|$.
3. Give the complete arguments showing how the contraction mapping theorem is used in the proof of Theorem 8.2.
4. Complete the proof of Theorem 8.3.

## 8.2 STABILITY AND CONVERGENCE

The results on the relation between stability, consistency, and convergence developed in Section 4.3 can be extended to the nonlinear case. Consider the discretized version of (8.1)

$$P_n x_n = 0, \tag{8.7}$$

where $x_n$ is an element of a finite-dimensional space $X_n$ and $P_n : X_n \to X_n$.

We will use the notation and the definitions introduced in Section 4.3. An approximation method is defined as a sequence of triplets $\{P_n, r_n, p_n\}$, where $r_n$ and $p_n$ are the restriction and prolongation operators previously defined. A method is said to be consistent of order $k$ if there exists a $c(x)$ such that

$$\|P_n r_n x - r_n P x\| \le c(x) n^{-k}$$

for all $x \in X$.

**DEFINITION 8.2.** The approximation method $\{P_n, r_n, p_n\}$ is said to be stable near $x \in Z$ if, for sufficiently large $n$, there exist constants $\rho$ and $c$ (independent of $n$) such that

$$\|z_n - \bar{z}_n\| \le c \|P_n z_n - P_n \bar{z}_n\| \tag{8.8}$$

for all $z_n, \bar{z}_n \in b(r_n x, \rho)$.

The central Theorem 4.13 is then easily generalized as Theorem 8.4.

**THEOREM 8.4.** If $\{P_n, r_n, p_n\}$ is a stable and consistent approximation of $P$ near $x \in Z$ and if (8.7) has a solution in $b(r_n x, \rho)$, then $x_n$ converges to $x$ and the order of convergence is at least as large as the order of consistency.

PROOF.   From (8.1) we have

$$r_n P x = 0,$$

so that

$$P_n x_n - P_n r_n x = r_n P x - P_n r_n x.$$

Since by assumption, $x_n$ is in $b(r_n x, \rho)$ we can apply (8.8) and

$$\|x_n - r_n x\| \le c \|r_n P x - P_n r_n x\| = O(n^{-k}),$$

where $k$ is the order of consistency.                                   ∎

In the preceding theorem we carefully chose our definitions and assumptions to make Theorem 8.4 appear as a simple generalization of Theorem 4.13. It is worthwhile to elaborate. First, stability as given in Definition 8.2 is not easily verified, but can be recast along the lines of Theorem 8.1.

**THEOREM 8.5.** Assume that for sufficiently large $n \geq n_0$ the operators $P_n$ are twice Fréchet differentiable at $r_n x$ with

$$\sup_{n \geq n_0} \| P_n'(r_n x)^{-1} \| \leq M \tag{8.9}$$

$$\sup_{n \geq n_0} \| P_n''(r_n x) \| \leq M. \tag{8.10}$$

Then the approximation method $\{ P_n, r_n, p_n \}$ is stable near $x \in Z$.

**PROOF.** Let $z_n, \bar{z}_n \in b(r_n x, \rho)$. Since

$$P_n'(z_n) = P_n'(r_n x) + P_n''(r_n x)(z_n - r_n x) + \eta(z_n, r_n x),$$

it follows by standard arguments that $\sup \| P_n'(z_n)^{-1} \|$ is bounded. Therefore, since

$$P_n z_n - P_n \bar{z}_n = P_n'(z_n)(z_n - \bar{z}_n) + \eta_1(z_n, \bar{z}_n),$$

(8.8) follows. ∎

We also assumed in Theorem 8.4 that $P_n x_n = 0$ has a solution in $b(r_n x, \rho)$. This is actually implied in the assumptions that were made.

**THEOREM 8.6.** Assume (8.9) and (8.10) are satisfied. Then for sufficiently large $n$ the equation $P_n x_n = 0$ has a unique solution in some neighborhood $b(r_n x, \rho)$.

**PROOF.** Consider

$$G_n z_n = z_n - P_n'(r_n x)^{-1} P_n z_n,$$

then proceed as in Theorem 8.2 to show that $G_n$ is a contraction mapping. ∎

While superficially these results for the nonlinear case look very much like those of the linear case there arise considerable practical difficulties, in that the checking of stability requires establishing a bound on $\| P_n'(r_n x)^{-1} \|$, which depends on the unknown solution $x$. This creates problems which are not easily dealt with in a general way. Nevertheless, for specific cases and algorithms it is possible to overcome the difficulty and to prove stability. For further discussion on this and for some examples see (Keller, 1975). Whether one can construct a general theory for this is an open question which is well worth pursuing.

This approach to stability analysis is essentially due to Keller (see the reference cited above). For a related, but slightly different analysis, see (Tucker, 1969). We chose Keller's method because of its similarity to the linear case.

## 8.3 ASYMPTOTIC EXPANSION OF THE DISCRETIZATION ERROR

In order to obtain asymptotic error estimates and to justify the various extrapolation techniques we now look at the possibility of an asymptotic error expansion. Again, the results look similar to the linear case with some technical complications arising from the nonlinearity.

As in Section 4.4 we say that the consistency error has an asymptotic error expansion of order $p$ at $x$ if there exists an operator $M:X\rightarrow Y$ such that

$$P_n r_n x = r_n P x - n^{-p} r_n M x + o(n^{-p}). \qquad (8.11)$$

The existence of an asymptotic expansion for the consistency error implies the existence of a similar expansion for the discretization error.

**THEOREM 8.7.**   Assume that

(a)   $\{P_n, r_n, p_n\}$ is a stable approximation to $P$ near $x \in Z$.

(b)   The consistency error has an asymptotic expansion of the form (8.11) at all points $x + \alpha e$, where $e \in X$ is the unique solution of $P'(x)e = Mx$.

(c)   $P$ and $M$ are sufficiently differentiable.

Then

$$x_n - r_n x = n^{-p} r_n e + o(n^{-p}) \qquad (8.12)$$

PROOF.   Since the method is stable we know that

$$\|x_n - r_n x\| = O(n^{-p}).$$

Also

$$P_n r_n x - r_n P x = P_n r_n x - P_n x_n$$
$$= -P_n'(r_n x)(x_n - r_n x) + o(n^{-p})$$

so that

$$P'_n(r_n x)(x_n - r_n x) = n^{-p} r_n M x + o(n^{-p}).$$

Consider now

$$P'_n(r_n x)(x_n - r_n x - n^{-p} r_n e) = P'_n(r_n x)(x_n - r_n x) - n^{-p} P'_n(r_n x) r_n e$$
$$= n^{-p} r_n M x - n^{-p} P'_n(r_n x) r_n e + o(n^{-p}).$$

$$(8.13)$$

Applying (8.11) to $x + n^{-p} e$ we have

$$P_n r_n(x + n^{-p} e) = r_n P(x + n^{-p} e) - n^{-p} r_n M(x + n^{-p} e) + o(n^{-p})$$

and by a Taylor expansion

$$n^{-p} P'_n(r_n x) r_n e = n^{-p} r_n P'(x) e - n^{-2p} r_n M'(x) e + o(n^{-p}). \quad (8.14)$$

From Assumption (b) we get

$$r_n P'(x) e = r_n M x,$$

and putting this into (8.14) and (8.13)

$$P'_n(r_n x)(x_n - r_n x - n^{-p} r_n e) = o(n^{-p}).$$

Since the method is assumed to be stable (8.12) follows. ∎

The reader may quite rightly feel a little uneasy about this proof because of the apparently indiscriminate use of $o(n^{-p})$ terms. This is a valid objection; arguments of this nature have to be used with extreme caution and instances of incorrect proofs of this type are not unheard of. Actually, in this case everything can be made precise and we urge the reader to do so.

A more extensive discussion of asymptotic error expansions and their application to Richardson's extrapolation can be found in (Stetter, 1965). A similar analysis in connection with the method of deferred corrections is presented in (Pereyra, 1966).

# 9

# THE SOLUTION OF IMPROPERLY POSED PROBLEMS

We have so far considered only problems which were well-posed. It seems intuitively reasonable that equations arising from physical situations should be well-posed since the parameters entering the equations are usually derived from actual measurements and therefore inexact to some extent. It would be unreasonable to expect the physical situation to change drastically if we were to change the parameters by a small amount; if our equation does indeed model the real world, then we would expect it to be insensitive to experimental errors. This reasoning is fairly sound and most of the problems we deal with are indeed well-posed. Nevertheless, certain mathematical descriptions of actual situations lead to improperly or ill-posed formulations, as we shall see shortly by a simple example.

The prototype for this situation is the *Fredholm equation of the first kind*

$$\int_0^1 k(s,t)x(t)\,dt = y(s), \qquad 0 \le s \le 1. \tag{9.1}$$

If we assume that $k(s,t)$ is differentiable in $[0,1]\times[0,1]$, then it is easy to see that this problem in not well-posed. Take the function $x_n(t) = \sin nt$; then

$$\int_0^1 k(s,t)x_n(t)\,dt = O(1/n),$$

which follows from an elementary argument. This according to Theorem 4.7, shows that the operator does not have a bounded inverse. Another way of saying this is that a small perturbation of size $O(1/n)$ on the right-hand side can generate a change with norm of order unity in the solution. Since we can make $n$ as large as we want the problem cannot be well-posed. Of course, we can write down any equation we want; it does not follow that there is an actual physical situation modeled by such an equation. For (9.1), however, there is a corresponding physical problem.

Consider a simple optical system consisting of an object $O$, a lens $L$, and an image $I$. If we denote by $x(s)$ and $y(s)$ the light intensities of $O$ and $I$, respectively, then we can think of the lens as represented by an operator $L$ mapping $x$ into $y$, or $Lx = y$. What form might this operator $L$ have? If the lens were perfect, then the image intensity at some point $s$ would depend only on the intensity at a single corresponding point in the object; if the lens is not perfect, then a certain amount of "smearing" will occur and $y(s)$ will depend on some sort of weighted average of the object intensity. Mathematically, this weighted average can be represented as an integral with a certain weight function; if we call this weight function $k(s,t)$, then we get exactly (9.1).

We may note at this point that if we take the Dirac delta "function" as $k(s,t)$, then $x(s) = y(s)$ (perfect lens case), and the problem is certainly well-posed. On the other hand, as we have indicated, when $k(s,t)$ is smooth then the problem is ill-posed. We are thus led to the somewhat unexpected situation that a poorly behaved kernel defines a solution more satisfactorily than a smoother kernel.

While we have taken a somewhat oversimplified physical situation in this discussion, equations of this type often arise in physical systems where the nature of an object is to be inferred from indirect and experimental observations. For a discussion of other types of ill-posed problems arising in mathematical physics see (Lavrentiev, 1967).

When solving ill-posed problems numerically we must certainly expect some difficulties, since any errors act as a perturbation on the original equation and so may cause arbitrarily large changes in the solution. Since errors can never be completely avoided the prospects look rather dim; if we are to make any progress we must reexamine our notion of the solution of an equation. To provide some insight into this, let us return to the simple physical model above. Suppose that the light intensity in the object oscillates rapidly from dark to bright. When this input is smeared by an imperfect lens the image will look uniformly grey. Therefore, if we observe a uniformly grey image it may be the result of an object composed of rapidly alternating dark and bright spots. On the other hand, it may also be the image of a uniformly grey object. Without any further assumptions we cannot tell the difference. If we were to encounter this in an actual experimental situation we would probably rule out the first alternative as uninteresting or implausible and ask instead what regularly behaved object could have caused the observed image.

This gives us a clue for proceeding in general: to solve ill-posed problems we must restrict the type of answer we will accept so that we get only plausible solutions. How we define "plausible" is essentially up to us; different ways of definition lead to different numerical approaches as we see below. Once we have defined an acceptable solution we still have some worries. How do we know that the problem actually has a solution in the defined set? The answer is that it probably has not, but if we are in a real situation the presence of experimental and observational errors makes it unimportant to satisfy the equation exactly. The task of solving an ill-posed problem can then be restated as: find an element $x$, subject to certain restrictions, such that the equation $Lx = y$ is satisfied approximately, that is, such that

$$\|Lx - y\| \leq \epsilon, \tag{9.2}$$

for some given $\epsilon > 0$.

For the rest of this chapter we will not proceed in the usual theorem-proof fashion that we have employed so far; rather we will present a "discussion" of the pertinent problems and results. By this we mean that we will indicate the lines along which a theory can be developed without rigorously carrying out all the proofs. The reader may consider it a challenge to his understanding to see how far he can make precise what we are about to present. The reason for taking this approach is two-fold. First, it seems that at the present there has not yet emerged a commonly accepted approach and it is difficult to see what a general theory will ultimately look like. Also, the theoretical development that does exist involves technical difficulties whose discussion would lead us too far afield. As a second consideration, the reader may find it instructive to see the kind of "plausibility" arguments that are often made in the early stages of research. One often tries to get a "feeling" for the subject before settling down to the task of establishing proofs and a general theory. One must be cautious, however. While this type of treatment can be very helpful to one's understanding, one must not jump to the conclusion that anything has been proved.

## 9.1  REGULARIZATION TECHNIQUES

We shall restrict ourselves to equations of the form

$$Lx = y, \tag{9.3}$$

where $L$ is a linear operator between two linear spaces $X$ and $Y$. We also assume that $\mathcal{D}(L) = X$ and $\mathcal{R}(L) = Y$, so that the problem has a solution.

Since we want to consider ill-posed problems we shall assume that $L^{-1}$ is not bounded.

In our discussion of (9.1) we saw that a difficulty was caused by the existence of highly oscillatory functions which, while not actually being solutions, satisfy the equation very closely. Somehow, we want to eliminate such "near-solutions" from consideration. One way to accomplish this is the so-called *regularization method*, which was first systematically developed by Tikhonov (see for example, Tikhonov, 1963). Consider the functional

$$J(x) = \|Lx - y\| + \alpha\varphi(x), \tag{9.4}$$

where $\alpha > 0$ and $\varphi$ is a functional defined on $X$. Instead of attempting to solve (9.3) we try to minimize $J$, that is, we look for an $\hat{x} \in X$ such that

$$J(\hat{x}) \le J(x) \text{ for all } x \in X \tag{9.5}$$

Provided that the *regularization functional* $\varphi(x)$ is properly chosen this minimization problem will be well-posed even if (9.3) is not. How then do we choose $\varphi$? Intuitively, we want $\varphi(x)$ to be small for well-behaved $x$ and very large for irregular $x$. More formally, we might require that $\varphi$ satisfy

(a) $\varphi(x) \ge 0$ for all $x \in X$
(b) $\varphi$ is such that for any $M < \infty$ the set $\Phi_M = \{x \in X | J(x) \le M\}$ is compact.

**Example 9.1.** Let $X$ be $C^{(1)}[0, 1]$ with maximum norm. Let $L$ be given by (9.1) and define $\varphi$ by

$$\varphi(x) = \|x\|_\infty + \|x'\|_\infty. \tag{9.6}$$

Condition (a) is obviously satisfied. Condition (b) follows from the fact that the functions in $\Phi_M$ are uniformly bounded and equicontinuous; hence by the Arzela–Ascoli theorem $\Phi_M$ is compact.

The reason for imposing conditions (a) and (b) is that they are sufficient to allow us to look for the minimum of $J$ on a compact set, which is a well-posed problem. We can see this from the following argument. If $x$ is a solution of (9.3), then $J(x) = \alpha\varphi(x)$. If $\hat{x}$ minimizes $J$, then

$$J(\hat{x}) \le J(x) = \alpha\varphi(x). \tag{9.7}$$

Thus the minimum, if it is attained, must be in the compact set $\{\bar{x} \in X | J(\bar{x}) \le \alpha\varphi(x)\}$.

Next, we consider the selection of $\alpha$ which controls the overall magnitude of the regularization effect. From one point of view we would like to make $\alpha$ as large as possible. Since for $\alpha = 0$ the problem is ill-posed it seems reasonable to require $\alpha$ to be large. Although the set in which the minimum lies is compact for all $\alpha > 0$, we run into ill-conditioning in the numerical minimization when $\alpha$ is very small. To see this, let us try to define a condition number for the minimization of a given functional $H(x)$. A condition number is essentially a measure of how well we can tell the solution from a near-solution; thus if we perturb the solution $x$ by $\Delta x$ we would like to see an appreciable change in $H(x)$. By a Taylor expansion we have

$$H(x + \Delta x) \cong H(x) + H'(x)\Delta x + \tfrac{1}{2} H''(x)(\Delta x)^2. \qquad (9.8)$$

If $x$ minimizes $H$, then $H'(x) = 0$, so that the perturbation $\Delta x$ causes a change of $\tfrac{1}{2} H''(x)(\Delta x)^2$ in $H$. Thus we can think of

$$1 / \inf \frac{\|H''(x)(\Delta x)^2\|}{\|\Delta x\|^2} \qquad (9.9)$$

as a condition number which we want to keep small. Since $J''$ involves a term $\alpha\varphi''$ it appears that a large $\alpha$ will tend to make the minimization of $J$ well-conditioned.

Of course, we also have to remember that minimizing $J$ does not yield the solution of (9.3) unless $\alpha = 0$, so for this reason we want to keep $\alpha$ small. If $x$ is a solution of (9.3) and $\hat{x}$ minimizes $J$, then we have

$$\begin{aligned} J(\hat{x}) &\leq J(x) \\ &= \|Lx - y\| + \alpha\varphi(x) \\ &\leq \alpha\varphi(x), \end{aligned}$$

from which we get

$$\begin{aligned} \|L\hat{x} - y\| + \alpha\varphi(\hat{x}) &\leq \alpha\varphi(x) \\ \|L\hat{x} - y\| &\leq \alpha\varphi(x). \end{aligned} \qquad (9.10)$$

Thus, while $\hat{x}$ is not a solution it will be a good approximate solution if $\alpha$ is small. If we are permitted an error $\epsilon$ in the sense of (9.2), then we must have $\alpha\varphi(x) \leq \epsilon$. A reasonable choice of $\alpha$ appears then to be

$$\alpha = \epsilon / \varphi(x), \qquad (9.11)$$

although because of the various inequalities this may not be the optimal choice.

Conditions (a) and (b) which we imposed on $\varphi$ were chosen, somewhat arbitrarily, for the sake of discussion. Other types of regularization functionals can and have been used. The proper choice of $\varphi$ obviously depends on $L$; furthermore we would like to choose it so that $\varphi(x)$ is very small when $x$ is the sought solution and so that it is large for those near-solutions we want to eliminate. The more information we have about the solution the better we will be able to do this. What the formal properties of $\varphi$ should be is an open question.

## 9.2  EXPANSION METHODS

In the regularization method we control the unwanted near-solutions by imposing a heavy penalty on them through the functional $\varphi(x)$. Another approach is to eliminate these undesirable solutions altogether by an *a priori* choice of the space of the permitted solutions. For this we choose a finite-dimensional space $X_n$ and look for solutions in this space; generally there will, of course, not exist an exact solution, so that we can solve (9.3) only approximately. If $\{\varphi_{ni}\}$ is a basis for $X_n$, then we look for a solution of the form

$$x_n = \sum_{i=1}^{n} \alpha_i \varphi_{ni} \qquad (9.12)$$

such that

$$Lx_n \cong y. \qquad (9.13)$$

This is essentially what we did in Section 4.2 for well-posed problems, although here we are not able to prove convergence since this concept is not very meaningful when the problem is ill-posed. In any case the observation suggests that it is possible to adapt Galerkin and least squares type methods for ill-posed problems.

Let us assume that $X$ and $Y$ are Hilbert spaces. If we use the Galerkin approach, then we want to make $Lx_n - y$ orthogonal to span $\{\varphi_{n1}, \varphi_{n2}, \ldots, \varphi_{nn}\}$, which leads as before to the matrix equation

$$A\alpha = \mathbf{b}, \qquad (9.14)$$

with $a_{ij} = (L\varphi_{ni}, \varphi_{nj})$, $b_i = (\varphi_{ni}, y)$.

The solvability of this system depends on the choice of $\{\varphi_{ni}\}$. One way which is, at least formally, very simple is to use the eigenfunctions of $L$ as expansion functions. If $u_i$ is such that

$$Lu_i = \lambda_i u_i,$$
(9.15)

then setting $\varphi_{ni} = u_i$ makes the matrix in (9.14) diagonal and the solution is immediately given by

$$x_n = \sum_{i=1}^{n} \frac{1}{\lambda_i} (u_i, y) u_i.$$
(9.16)

If we assume that the eigenfunctions of $L$ are closed in $X$, then formally we can write

$$x = \sum_{i=1}^{\infty} \frac{1}{\lambda_i} (u_i, y) u_i$$
(9.17)

and

$$\|x - x_n\|^2 \leq \sum_{i=n+1}^{\infty} \frac{1}{\lambda_i^2} (u_i, y)^2.$$
(9.18)

Whether (9.17) actually defines a convergent series depends on $y$ as well as on the distribution of the eigenvalues of $L$. If the series does converge, then (9.18) apparently indicates that we can get arbitrarily high accuracy by making $n$ sufficiently large. Furthermore, since the computation of $x_n$ by (9.16) seems to be a well-conditioned numerical process one might think that this gives us a way of solving ill-posed problems with no difficulty whatsoever! Such a conclusion ought to look very suspicious and there is indeed a flaw in our reasoning. The eigenvalues and eigenfunctions of $L$ are generally not known and must be computed numerically. This means that the eigenvalues of small magnitude and their associated eigenvectors will have large relative errors, making it difficult to evaluate (9.16) accurately. Therefore, one must try and keep $n$ as small as possible within the constraints of accuracy requirements.

Eigenfunction expansions for ill-posed problems (specifically the Fredholm equation of the first kind) were suggested in (Baker et al., 1964). More recent treatments advocate the more general singular function expansions. If $\{\lambda_i, u_i, v_i\}$ satisfy

$$Lu_i = \lambda_i v_i, \quad L^* v_i = \lambda_i u_i$$
(9.19)

where $L^*$ denotes the adjoint of $L$, then the $\lambda_i$ are called the singular values and $u_i, v_i$ the singular functions of $L$. If we modify Galerkin's method so that $x_n$ is a linear combination of the $u_i$, but $Lx_n - y$ is made orthogonal to span$\{v_i\}$, then the above analysis goes through in essentially the same way.

Of course it is not necessary to use eigenfunctions or singular functions in the expansion; any set closed in $X$ will do. The major advantage of the eigenfunction expansion is that the analysis of the conditioning of (9.14) and error bounds, such as (9.18), are easy to obtain.

Also, as an alternative to Galerkin's method we can use a least-squares technique, leading to a system of the form (9.14) with $a_{ij} = (L\varphi_{ni}, L\varphi_{nj})$ and $b_i = (L\varphi_{ni}, y)$. A considerable amount of work on the use of least-squares methods in this setting has been done (see Kammerer and Nashed, 1972). Another avenue of approach uses the fact that ill-posed problems arise primarily in connection with the "unravelling" of experimental observations as indicated at the beginning of this chapter. Since the observational errors often have known probability distributions it is possible to satisfy (9.3) subject to certain statistical requirements. This idea is explored in (Strand and Westwater, 1968).

Finally, we caution the reader not to take our rather descriptive treatment of the subject as an indication that there are no theoretical treatments of ill-posed problems. (Although it seems fair to say that there exists no unified theory at the present.) The already mentioned work of Kammerer and Nashed and the contributions of Wahba [see for example, (Wahba, 1973)] use sophisticated function-analytic methods to explore these questions. The regularization methods have been studied extensively in Russia; a large number of papers on the topic can be found in the Russian literature, particularly in the U.S.S.R. journal *Computational Mathematics and Mathematical Physics*.

# REFERENCES

Ahlberg, J. H., E. N. Nilson, and J. L. Walsh. *The Theory of Splines and their Application.* Academic, New York, 1967.

Anselone, P. M. *Collectively Compact Operator Approximation Theory and Applications to Integral Equations.* Prentice-Hall, Englewood Cliffs, NJ, 1971.

Aubin, J. P. *Approximation of Elliptic Boundary-Value Problems.* Wiley-Interscience, New York, 1972.

Baker, C. T. H., L. Fox, D. F. Mayer, and K. Wright. Numerical Solution of Fredholm Integral Equations of the First Kind, *Comput. J.*, 7, 141–148 (1964).

Birkhoff, G. and C. R. DeBoor. Piecewise Polynomial Interpolation and Approximation, in *Approximation of Functions*, H. L. Garabedian, Ed. American Elsevier, New York, 1965, 164–190.

Céa, J. *Optimisation Théorie et Algorithmes.* Dunod, Paris, 1971.

Chatelin, F. Convergence of Approximation Methods to Compute Eigenelements of Linear Operations, *SIAM J. Numer. Anal.* 10, 939–948 (1973).

Daniel, J. W. *The Approximate Minimization of Functionals*, Prentice-Hall, Englewood Cliffs, NJ, 1971.

Davis, P. J., *Interpolation and Approximation.* Blaisdell, New York, 1963.

Davis, P. J. and P. Rabinowitz. *Numerical Integration.* Blaisdell, Waltham, MA, 1967.

Graves, L. M. *The Theory of Functions of Real Variables*, 2nd ed. McGraw-Hill, New York, 1956.

Isaacson, E. and H. B. Keller. *Analysis of Numerical Methods.* Wiley, New York, 1966.

Kammerer, W. J. and M. Z. Nashed. Iterative Methods for the Best Approximate Solutions of Linear Equations of the First and Second Kinds, *J. Math. Anal. Appl.* 40, 547–573 (1972).

Kantorovich, L. V. and G. P. Akilov. *Functional Analysis in Normed Spaces* (translated from the Russian by D. E. Brown). Pergammon, Oxford, 1964.

Keller, H. B. *Numerical Methods for Two-point Boundary Value Problems.* Blaisdell, Waltham, MA, 1968.

Keller, H. B. Approximate Methods for Nonlinear Problems with Application to Two-point Boundary Value Problems, *Math. Comp.* 29, 464–474 (1975).

Lavrentiev, M. M. *Some Improperly Posed Problems of Mathematical Physics* (translated from the Russian by R. J. Sacker). Springer, Berlin, 1967.

Linz, P. A General Theory for the Approximate Solution of Operator Equations of the Second Kind, *SIAM J. Numer. Anal.* **14**, 543–554 (1977).

Mikhlin, S. G. *The Numerical Performance of Variational Methods* (translated from the Russian by R. S. Anderssen). Wolters-Noordhoff, Groningen, 1971.

Noble, B. Error Analysis of Collocation Methods for Solving Fredholm Integral Equations, in *Topics in Numerical Analysis*, J. J. Miller, Ed. Academic, London, 1973, 211–232.

Nyström, E. J. Über die Praktische Auflösung von Integralgleichungen mit Anwendung auf Randwertaufgaben, *Acta Math.* **54**, 185–204 (1930).

Pereyra, V. On Improving an Approximate Solution of a Functional Equation by Deferred Corrections, *Numer. Math.* **8**, 376–391 (1966).

Phillips, J. L. The Use of Collocation as a Projection Method for Solving Linear Operator Equations, *SIAM J. Numer. Anal.* **9**, 14–28 (1972).

Prenter, P. M. *Splines and Variational Methods*. Wiley-Interscience, New York, 1975.

Rall, L. B. *Computational Solution of Nonlinear Operator Equations*. Wiley, New York, 1969.

Rice, J. R. *The Approximation of Functions*, Vol. 2. Addison-Wesley, Reading, MA, 1969.

Richtmyer, R. D. and K. W. Morton. *Difference Methods for Initial Value Problems*, 2nd ed. Wiley-Interscience, New York, 1967.

Rivlin, T. J. *An Introduction to the Approximation of Functions*. Blaisdell, Waltham, MA, 1969.

Rudin, W. *Real and Complex Analysis*. McGraw-Hill, New York, 1966.

Rudin, W. *Principles of Mathematical Analysis*, 3rd ed. McGraw-Hill, New York, 1964.

Schultz, M. H. *Spline Analysis*. Prentice-Hall, Englewood Cliffs, NJ, 1972.

Stetter, H. J. Asymptotic Expansions for the Error of Discretization Algorithms for Nonlinear Functional Equations, *Numer. Math.* **7**, 18–31 (1965).

Strand, O. N. and E. R. Westwater. Minimum-RMS Estimation of the Numerical Solution of Fredholm Integral Equations of the First Kind, *SIAM J. Numer. Anal.* **5**, 287–295 (1968).

Strang, G. and G. J. Fix. *An Analysis of the Finite Element Method*. Prentice-Hall, Englewood Cliffs, NJ, 1973.

Taylor, A. *Introduction to Functional Analysis*. Wiley, New York, 1964.

Tikhonov, A. N. Solution of Incorrectly Formulated Problems and the Regularization Method, *Sov. Math. Dokl.* **4**, 1036–1038 (1963).

Tucker, T. S. Stability of Nonlinear Computing Schemes, *SIAM J. Numer. Anal.* **6**, 72–81 (1969).

Wahba, G. A Class of Approximate Solutions to Linear Operator Equations, *J. Approx. Theory* **9**, 61–77 (1973).

Wilkinson, J. H. *Rounding Errors in Algebraic Processes*, Prentice-Hall, Englewood Cliffs, NJ, 1963.

# INDEX